BUILDING NEURAL NETWORKS FROM SCRATCH WITH PYTHON

L.D. KNOWINGS

CONTENTS

INTRODUCTION - BUILDING NEURAL NETWORKS FROM SCRATCH WITH PYTHON

Has it ever occurred to you how a machine can recognize images, understand language, and compete and defeat human experts in complex games? Well, the answer to that question lies within neural networks.

These networks, also known as artificial or simulated neural networks (ANN or SNN), are a subset of machine learning. It's important to know that neural networks are paramount since they enable machine learning algorithms to generate human-like results.

Their ability to mimic complex functionalities of the human brain is what gives them their name and structure. Given the rapid technological advancements in AI and machine learning, I can confidently tell you need to understand how these neural networks work and how you can create one.

Neural networks have become so essential that their market value of just $14.35 billion in 2020 is now expected to experience a capital annual growth rate (CAGR) of around 26.7%. This means that the estimated value of the neural networks market will reach $152.61 billion by 2030.

So, if you want to cut yourself a piece of the pie and make strides in the software industry, you must know how to build neural networks with Python.

The first thing that you need to know is that building a neural network can indeed be challenging. However, this doesn't mean that it's impossible. Over the years, I've often found that people find this to be a challenge for several reasons, and before we continue any further, I'll share those reasons with you.

One of the most significant difficulties I've seen people face is understanding complex concepts and algorithms like backpropagation, ReLU, and reinforcement learning. Before we start, I want you to know that the theories and algorithms associated with developing neural networks can be complex for some, especially if they've been out of the *"learning"* mindset for some time.

Believe me when I tell you that I've seen countless people give up on their dream of developing a neural network because they got overwhelmed with the learning curve. However, the problem isn't the learning curve or the concepts. I've found that this information is not made accessible as well as it could be.

You need to understand that this inaccessibility of knowledge and guidance can make tackling challenges even more difficult. When you begin developing your neural network, you'll likely experience problems with coding and debugging. In such a situation, not having access to the proper knowledge and guidance can make things worse.

The inaccessibility to the required knowledge and the problems one encounters will likely deter one's ability to accomplish whatever one wants with neural networks. I've also seen many developers have a challenging time when it comes to explaining the decisions of neural networks.

It's important to know that this problem may occur for various reasons, and we'll go over them later in the book. However, the impact of this problem is that it makes it difficult for developers to determine if the work is *"good enough."* This leads to endless modifications using frameworks and techniques like TensorFlow and Dropout.

I want you to know that one might come from a STEM background, but it's still possible for them to experience any of these challenges or similar ones. Throughout this book, we'll focus on comprehensively understanding how you can overcome such challenges.

We'll focus on understanding how current and expected technology trends can influence your work with neural networks. In this book, you'll learn all there is to know about developing a neural network with Python. You'll be able to use this knowledge to boost your skill set and excel in the software industry like never before.

But that's not all. The benefits here are truly endless. You'll learn how neural networks can be applied in the real world. They're a powerful tool for data scientists and statisticians and have a wide range of real-world applications such as:

- Being used for image processing and facial recognition.
- Making real-time predictions about stock market movements.

- Modeling non-linear time dynamic systems used in Aerospace engineering.

Once you've fully understood how neural networks can be applied in the real world, we'll gain a comprehensive understanding of the foundations of a neural network. Then, we'll look at the math needed for neural networks.

Once we're done with these fundamental stages and the development, we'll look at how you can implement the neural network. This will involve things like importing libraries and data loading and preprocessing.

Remember when I mentioned that accessibility to information and knowledge is a common problem people face when working on a neural network?

That won't be the case here since I'll share some great troubleshooting tips I've also picked up over the years. You'll learn to tackle everything from vanishing gradients to underfitting and more.

I promise you that by the end of this book, you'll be more advanced, knowledgeable, and skillful than others in their careers. When you walk into the office, your co-workers will envy you, and you will be the one they look up to. Why? Because you'll be the one who can produce and employ their neural network using Python.

I'm a father and a track coach. I started in the information technology (IT) industry around 24 and now have nearly three decades of experience. After gaining substantial experience as an IT executive, I now run my own managed services company, helping businesses push the boundaries of technological excellence.

I genuinely understand that aspiring professionals in the neural networks space face significant challenges either because the information is not available or not accessible. And that right there is my aim for writing the book.

I want to do my part in eliminating this challenge and helping you and others like you excel because I know that dreams of developing a neural network can not be fulfilled without the proper knowledge and guidance.

So, with that in mind, let's get started!

1

INTRODUCTION TO NEURAL NETWORKS

"Once a new technology rolls over you, if you're not part of the steamroller, you're part of the road."

— STEWART BRAND, WRITER

This wise saying holds true in the ever-changing world of technology. With that in mind, let's simplify the complexity of neural networks, allowing you to dive headfirst into deep learning and artificial intelligence. By the end of this chapter, you will grasp the inner workings of neural networks, their practical applications, and the fundamental elements that constitute these algorithms.

NEURAL NETWORKS AND THEIR REAL-WORLD APPLICATIONS

So, what exactly is a neural network? It's a fundamental concept in artificial intelligence, a method designed to enable computers to process data in a manner inspired by the intricate workings of the human brain. Neural networks are a subset of machine learning, specifically deep learning, which relies on interconnected nodes or neurons organized in a layered structure, mirroring the neural connections in the human brain. These connections form an adaptive system that empowers computers to learn from their errors and continuously enhance their performance.

So basically, artificial neural networks are at the forefront of addressing complex problems, whether summarizing lengthy documents, recognizing faces in photographs, or tackling various other intricate tasks. They provide a ground-breaking strategy that combines the strength of technology with the cognitive agility of the human mind, resulting in unprecedented accuracy and efficiency.

Let's have a look at some real-world neural network applications.

Image Recognition and Computer Vision

In the real world, neural networks, particularly Convolutional Neural Networks (CNNs), serve as the backbone of various applications, with image recognition being a prominent example.

CNNs excel at image classification, the process of taking an input (such as an image) and producing an output, either a definitive class label (like "cat") or a probability score indicating the likelihood of the input belonging to a specific category (e.g., a 90%

probability of it being a cat). While humans can glance at a picture and instantly recognize familiar faces, the journey for a computer is quite different. It involves a neural network, more specifically, a convolutional neural network.

Think about how, in a split second, you discern whether a person passing by is a familiar acquaintance or a stranger. It's an intuitive process: you observe their unique characteristics, such as their face shape, eye color, hairstyle, body type, gait, or clothing, and match them to someone you know. This rapid cognitive feat is similar to what neural networks do, albeit more structured and systematic.

A system must first learn to identify these distinct features to recognize faces. Unlike traditional machine learning models, deep learning models, like CNNs, take a different approach. CNNs have revolutionized image recognition. They excel in analyzing visual data, working silently in the background, and can be found in applications ranging from Facebook's photo tagging to autonomous vehicles. They operate swiftly and efficiently, but how do they achieve this?

Neural networks have significantly elevated accuracy in image recognition thanks to deep learning advancements and extensive datasets. This technology finds application in self-driving cars, enabling them to recognize road signs and other road users, content moderation to automatically filter inappropriate content, facial recognition to identify individuals and their attributes, and image labeling to spot brand logos, clothing, and other visual details.

Natural Language Processing with Neural Networks

Natural language processing (NLP) relies heavily on neural networks, which enable a wide range of applications such as text classification, sentiment analysis, and machine translation. The importance of neural networks in natural language processing cannot be overstated. They have elevated the accuracy and performance of NLP tasks to unprecedented levels. Two primary architectural stars in NLP are Recurrent Neural Networks (RNNs) and Transformers, which are adept at handling sequential data.

RNNs are used in technologies that we use every day, from speech recognition on our smartphones to multilingual translation and even stock market predictions. These neural networks are born with the ability to model sequences, allowing robots to delve into the complexities of human language. What truly sets them apart is their capacity to predict words and tackle subjects not explicitly ingrained during their training.

To reach such proficiency in NLP, neural networks require extensive training on substantial text corpora tailored to the specific language or text type. In NLP language models, neural networks act early, transforming words in our vocabulary into continuous vectors. Their magic lies in understanding the meaning of a word concerning its neighboring words within a text. These word vectors facilitate intricate operations that yield semantically meaningful results.

Recommender Systems

Recommender systems have become pivotal in delivering tailored content and product recommendations across various domains, from streaming platforms to e-commerce and social media. These systems use neural networks to enhance their effectiveness, offering users a highly personalized experience.

Neural networks in recommender systems excel at understanding user preferences by learning from their interactions with items. They employ complex algorithms to capture intricate patterns and relationships within user-item data. This results in recommendations that align closely with individual tastes.

One key feature of neural network-powered recommender systems is user and item embeddings. These embeddings represent preferences and characteristics as continuous vectors. It allows the system to comprehend the nuances of user choices and item attributes. By optimizing model parameters through extensive training, neural networks minimize prediction errors, leading to more accurate recommendations.

These networks handle various recommendation scenarios, including item-based, personalized, and content-based approaches. This versatility ensures that users receive suggestions that resonate with their unique preferences and interests. What's more is that companies are exploring hybrid recommendation systems that integrate the strengths of content-based, collaborative filtering, and knowledge-based methods with neural networks. This innovative approach promises even more accurate and personalized recommendations, heralding a new era in recommender systems.

BASIC BUILDING BLOCKS OF NEURAL NETWORKS

Now that you've glimpsed the real-world applications of neural networks, it's time to delve into the fundamental building blocks that underpin this transformative technology. At the core of neural networks are two essential elements: neurons and layers.

Neurons serve as the fundamental processing units, and it's their activation functions that enable neural networks to make decisions and learn from data. On the other hand, layers organize these neurons into a structured hierarchy with intricate connections that allow information flow. Understanding these fundamental components is essential for deciphering the complicated world of neural networks and exploiting their ability to solve complex problems with precision and efficiency.

Neurons and Their Activation Functions

Neurons and their activation functions are the fundamental building blocks of neural networks. These elementary components are responsible for processing information, making decisions, and ultimately driving the primary force of deep learning.

Imagine neurons as the computational powerhouses of the neural network. They are pivotal in receiving inputs, conducting computations, and producing outputs. Here's how they operate:

- Neurons receive a set of inputs, each associated with a weight. The inputs are multiplied by their respective weights, introducing a significance level to each input.
- Once the weighted inputs are calculated, they are summed together with a bias term. This summation creates a

combined value that encapsulates the significance of various inputs.

- The weighted sum undergoes a crucial transformation through the activation function. This function imparts non-linearity to the network and ultimately determines the neuron's output.

Diverse Activation Functions

Activation functions are a vital element in shaping neural networks' capabilities. They introduce non-linear properties, allowing the network to learn intricate patterns and relationships.

Several activation functions are at the forefront of deep learning:

1. **Sigmoid or Logistic Activation Function:** This function produces an S-shaped curve and maps input values to a range between 0 and 1. It is beneficial for binary classification tasks where results must fall within this range.
2. **Tanh (Hyperbolic Tangent) Activation Function:** Tanh is similar to the sigmoid function but with a range between -1 and 1. It is often used in hidden layers, helping to center data and make learning in subsequent layers more efficient.
3. **ReLU (Rectified Linear Unit) Activation Function:** ReLU is the most widely used activation function known for its efficiency and simplicity. It sets negative values to zero, keeping positive values unchanged. It is a workhorse for many neural networks due to its computational advantages.

4. **Softmax Activation Function:** Softmax is employed for multi-class classification tasks. It provides probability scores for each class, allowing the network to make informed decisions regarding class membership.

Why Activation Functions Are Essential

Activation functions add a critical layer of non-linearity to the neural network. Without them, the network would essentially perform linear transformations on the inputs, making it incapable of learning complex tasks.

It would reduce the network to a mere linear regression model. Activation functions inject non-linearity, enabling the network to capture intricate relationships and perform the complex tasks that make neural networks so powerful.

The Role of Nodes

Now, let's step back and clarify a crucial concept: nodes. Nodes are analogous to neurons in our neural network and play a pivotal role. They receive input signals like neurons responding to external stimuli in our brains. In this context, the activation function decides whether these nodes should be activated, determining the importance of their input to the network's predictions.

In essence, activation functions are the cornerstone of neural networks, and they introduce an additional, yet indispensable, step during the forward propagation process. While this may seem like an extra computational burden, the value they add is immeasurable.

They empower the network to tackle complex problems, recognize patterns, and transform input data into meaningful and actionable output.

Layers and Their Connections

Neural networks are structured as a series of interconnected layers, each playing a unique role in learning. Understanding the basics of these layers and their connections is essential to comprehend how neural networks function.

To unravel the essence of neural networks, let's break it down step by step:

1. **Data Influx:** The journey begins by feeding input data into the neural network, marking the inception of information flow.
2. **Layer-to-Layer Transition:** The data flows systematically from one layer to the next until it reaches the output layer, shaping the outcome as it traverses these computational strata.
3. **Error Computation:** Once we have the output, the network calculates the error, distilling the model's performance into a scalar value.
4. **Parameter Adjustment:** Here lies the crux of learning. To improve the network's performance, we adjust specific parameters, such as weights and biases, by subtracting the derivative of the error concerning the parameter itself.
5. **Iterative Process:** This process is not static; it's an iterative journey of refining and enhancing the network's capabilities.

The Need for Modular Layer Design

The pivotal question is: How do we ensure that our neural network can adapt to different architectures, activation functions, and loss functions while maintaining its computation efficiency?

The answer lies in a modular approach. Every layer in a neural network should be implemented separately to enable flexibility. Regardless of the network's architecture, activation functions, or the type of loss used, it should be able to compute derivatives effectively. This modularity is the cornerstone of a versatile and adaptable neural network.

The Role of Layers and Nodes

Now, let's explore a fundamental aspect of configuring a neural network: determining the number of layers and the number of nodes in each hidden layer. These hyperparameters govern the architecture of the network and are crucial in tailoring the model to your specific predictive modeling problem.

A node, also known as a neuron or perceptron, serves as a computational unit in a neural network. It receives one or more weighted input connections, processes them through a transfer function, and generates an output connection. Nodes are organized into layers to form a network.

Why Have Multiple Layers?

The choice of how many layers to specify is vital, but understanding why we opt for multiple layers is equally essential. A single-layer neural network can only represent linearly separable functions. A single layer suffices for simple problems where data can be divided by a line. However, most real-world problems are not linearly separable. This is where a Multilayer Perceptron (MLP) comes into play.

An MLP can represent convex regions, effectively learning to draw shapes around high-dimensional data to separate and classify it. A key finding by Lippmann in 1987 states that an MLP with two hidden layers is adequate for creating classification regions of any desired shape.

How Many Layers and Nodes to Use?

The challenge emerges in determining your neural network's precise number of layers and nodes. There's no analytical formula to calculate these hyperparameters. In essence, you are embarking on uncharted territory. No one has solved your problem before you, so no one can provide a definitive answer.

The number of layers and nodes in each layer are model hyperparameters that must be specified. The configuration must be discovered through rigorous testing and controlled experiments. This involves systematic experimentation to understand how best to structure your network for your unique predictive modeling problem.

Intuition plays a significant role in configuring your neural network. Your understanding of the problem domain may lead you to believe that a deep hierarchical model is necessary. A deep network provides a hierarchy of layers that build up levels of abstraction, offering a deeper insight into your data. Intuition, combined with systematic experimentation, is the key to unlocking the optimal configuration of layers and nodes for your neural network.

UNDERSTANDING THE STRUCTURE AND COMPONENTS OF A NEURAL NETWORK

Neural networks, the powerhouse behind modern AI, are fascinatingly intricate yet remarkably intuitive once you grasp their fundamental structure and components. Let's delve into the critical elements of a neural network: the input, hidden, and output layers.

Input Layer

- The input layer serves as the entry point for data into the neural network. It represents the features, variables, or data points to be processed.
- Each neuron in the input layer corresponds to a specific feature or input dimension. For example, each neuron may represent a pixel value in image recognition.
- The input layer passes the raw data to the subsequent layers for processing.

Hidden Layer

- Hidden layers are the heart of a neural network. They perform the bulk of the computation required for learning and making predictions.
- A neural network may contain one or more hidden layers, depending on its complexity and task. Deep neural networks have multiple hidden layers.
- Neurons in the hidden layers process the input data using weighted connections and activation functions. The weights determine the strength of the connections, and the

activation functions introduce non-linearity, enabling the network to model complex patterns and relationships.

- The hidden layers transform the input data through mathematical operations, gradually abstracting and extracting features.

Output Layer

- The output layer is the final segment of a neural network. It produces the network's predictions based on the computations carried out in the hidden layers.
- The number of neurons in the output layer depends on the specific task the network is designed for. One neuron is usually in binary classification, while multiclass classification or regression tasks may require multiple neurons.
- The output layer's neurons generate the network's predictions, representing class probabilities, continuous values, or categorical labels.

Feedforward and Backpropagation

Understanding the inner workings of feedforward and backpropagation is paramount in the universe of neural networks. These concepts are the pillars upon which artificial neural networks rest, enabling them to perform complex tasks efficiently.

Feedforward neural networks are like well-orchestrated symphonies where information flows in a harmonious, unidirectional manner without feedback loops. This type of neural network, often called a multi-layer neural network, consists of input, hidden, and output layers. It's through these layers that information takes its transformative journey.

How Feedforward Neural Networks Operate

Input nodes are the entry points where data is received. This data embarks on a journey through the network, going through hidden layers before reaching the output nodes. Crucially, no links exist within the network to allow information to travel backward from the output nodes. This unidirectional flow ensures that information only proceeds in one direction.

Feedforward neural networks prove invaluable across diverse domains, including image recognition, natural language processing, classification, regression, and pattern recognition. They are the workhorses behind the scenes, powering many applications.

Backpropagation

Backpropagation, an indispensable facet of neural network training, fine-tunes the network's weights based on the error rate obtained during forward propagation. This iterative refinement enhances the network's precision and generalization.

- Backpropagation is the essence of neural network training, ensuring that the weights are adjusted to minimize the cost function.
- By feeding the error rate backward through the neural network layers, backpropagation tunes the network's weights, biases, and activation functions.
- The gradients of the cost function dictate the extent of these adjustments, resulting in an ever-improving neural network.

In a Nutshell

Feedforward neural networks usher data along a one-way path, making them exceptional approximators of functions. They excel in object detection in images and classification tasks. These networks can be as straightforward as a single-layer perceptron or as complex as multi-layered perceptrons.

Backpropagation, on the other hand, refines the network's accuracy by minimizing error through iterative weight adjustments. This process often likened to fine-tuning, is central to neural network training and ensures a reliable, efficient, and highly generalizable model.

Together, these components form the structure of a neural network. The network's learning and decision-making process involves passing data from the input layer through the hidden layers to the output layer, gradually transforming and abstracting the information. Neural networks utilize algorithms like backpropagation to adjust the weights of connections during training, improving their ability to make accurate predictions.

Segue

In Chapter 2, we'll delve into the mathematical foundations of neural networks. You'll unlock the secrets of gradient descent, grasp the intricacies of backpropagation, and master the art of weight and bias adjustments, all while exploring diverse loss functions and optimization methods.

## 2

FOUNDATIONS OF NEURAL NETWORKS

"The advance of technology is based on making it fit in so that you don't really even notice it, so it's part of everyday life."

— BILL GATES

Neural networks, at their essence, are mathematical constructs. To truly master their workings, one must first wade through the mathematical underpinnings that govern their behavior. This section doesn't merely serve as a cursory introduction; it endeavors to illuminate the foundational mathematics that renders neural networks powerful and versatile.

From linear algebra to calculus, the mathematical paradigms we will explore here are pivotal in shaping neural networks from theoretical constructs to practical tools for complex problem-solving. By understanding this core mathematics, readers will be better equipped to harness the true potential of neural networks, making informed decisions in design, implementation, and optimization.

Dive in, and let's demystify the equations and notations that form the bedrock of this captivating domain.

UNDERSTANDING THE CORE MATHEMATICS OF NEURAL NETWORKS

Linear algebra plays a fundamental role in understanding and analyzing neural networks, which are quantitative models that adaptively learn to associate input and output patterns using learning algorithms. This chapter will explore key concepts from linear algebra essential for comprehending these models and their associated matrix operations.

It provides a robust framework for analyzing neural networks, particularly those known as associators. These models are designed to adaptively learn how to associate input and output patterns through learning algorithms. When the input patterns differ from the output patterns, these models are called heteroassociative.

Linear Algebra for Matrix Operations

Deep learning often begins with studying feedforward neural networks, which serve as fundamental building blocks in the field. These networks are essentially composite functions that involve multiplying matrices and vectors, a concept rooted in linear algebra.

In deep learning, vectors and matrices are crucial in performing various operations efficiently. Let's shed light on the importance of these mathematical constructs.

The Power of Vectors and Matrices

A feedforward neural network is a complex set of operations designed to propagate information forward. While the network might appear intricate, vectors and matrices are at its heart. These elements provide an efficient way to represent and process data.

Every column in the network corresponds to a vector. Vectors, in essence, are dynamic arrays containing data or features. The vector 'x' represents the input in our neural network example. While describing inputs as vectors is not mandatory, doing so simplifies parallel operations, a significant advantage in deep learning.

Deep learning, especially neural networks, can be computationally intensive, making it essential to employ optimization techniques. One such technique is vectorization, which dramatically speeds up computations. This is why GPUs, specialized in vectorized operations like matrix multiplication, are indispensable for deep learning.

Breaking Down the Equations

In the context of a feedforward neural network, the output of the hidden layer 'H' can be calculated using the equation:

$$H = f(W.x + b)$$

Here, 'W' represents the weight matrix, 'b' is the bias, and 'f' is the activation function. We don't need to go into the intricacies of feedforward neural networks. Understanding the role of matrices and vectors in these calculations is crucial.

Before diving into the multiplication of matrices, let's clarify some notations. Typically, vectors are represented by lowercase bold italic letters (e.g., x), while matrices are denoted by uppercase bold italic letters (e.g., X). If a letter is bold and capitalized but not italic, it represents a tensor.

From a computer science perspective:

- **Scalar:** A single number.
- **Vector:** A list of values (a rank 1 tensor).
- **Matrix:** A two-dimensional array of values (a rank 2 tensor).
- **Tensor:** A multi-dimensional matrix with rank 'n'.

From a mathematical perspective, a vector is a quantity with magnitude and direction. In a 2D space, vectors can be obtained through linear combinations of basis vectors (usually denoted as 'i' and 'j'). These basis vectors are unit normals with a one magnitude and are perpendicular to each other, making them linearly independent. This linear independence ensures that any vector in 2D space can be obtained through a linear combination of these basis vectors.

Matrices, on the other hand, are 2D arrays of numbers that represent transformations. Each column in a 2x2 matrix represents a basis vector after a transformation in 2D space. The identity matrix (denoted as 'I') is a unique matrix that doesn't alter the vector it operates on. The determinant of a matrix, det(A), represents the scaling factor of the linear transformation described by the matrix.

Understanding these mathematical perspectives is essential for deep learning researchers as they provide insights into the fundamental design concepts of key elements in deep learning.

Leveraging Libraries for Implementation

While comprehending these mathematical concepts is valuable, practical implementations often rely on libraries like NumPy for Python. NumPy simplifies working with arrays, matrices, and mathematical operations, making it an essential tool for deep learning in Python.

In NumPy, you can create arrays using 'np.array', generate random numbers with 'np.random', and perform matrix operations with the 'dot' method. This library streamlines complex mathematical operations, making it accessible for deep learning practitioners.

Exploring Types of Matrices

In the realm of linear algebra, various types of matrices serve distinct purposes:

1. **Diagonal Matrix:** This matrix has all zero elements except on the main diagonal.
2. **Identity Matrix:** An identity matrix is a special diagonal matrix with diagonal values set to 1.
3. **Symmetric Matrix:** A matrix is symmetric if it is equal to its transpose, represented as A = transpose(A).
4. **Singular Matrix:** Singular matrices have a determinant of zero, and their columns are linearly dependent. Their rank is less than the number of rows or columns.

Matrix Decomposition

Matrix decomposition, or matrix factorization, involves breaking down a matrix into a product of matrices. Various matrix decompositions are employed for specific problem classes. One widely used decomposition is eigen decomposition, representing a matrix as a set of eigenvectors and eigenvalues.

An eigenvector of a square matrix A is a non-zero vector 'v' for which multiplication by A only scales 'v' by a constant factor (λ). This relationship is defined as $A . v = lambda . v$, where 'v' is the eigenvector, and λ is the eigenvalue.

Eigen decomposition finds practical application in machine learning, particularly in dimensionality reduction.

The Role of Norms

Norms are essential in understanding machine learning concepts like regularization. In machine learning, overfitting and underfitting are common concerns.

Overfitting: This occurs when a model learns the training data too well, resulting in high training accuracy but low validation accuracy.

Underfitting: In contrast, underfit models struggle to learn the training data, leading to low training and validation accuracy.

Regularization techniques are employed to combat overfitting. Two prominent methods are L1 regularization (Lasso) and L2 regularization (Ridge). These methods rely on the concept of norms to control the size of model coefficients.

The L^2 norm (with p=2) is the Euclidean norm, representing the Euclidean distance between the origin and a vector. The L^1 norm is the sum of all vector elements, useful when precision is paramount. It helps distinguish between zero and non-zero elements and is called the Manhattan norm. Another norm of significance is the max norm, which is the absolute value of the element with the most significant magnitude. In matrices, the Frobenius norm serves as the L^2 norm equivalent.

Norms aren't just vital for regularization but also play a crucial role in optimization procedures within machine learning.

Vectorization and Broadcasting

Two vital concepts come into play to enhance the efficiency of deep learning computations: vectorization and broadcasting.

Vectorization involves rewriting loops to enable parallel execution by representing data as vectors. This technique takes advantage of CPUs' "vector" or "SIMD" instruction sets, allowing simultaneous execution of the same operation on multiple data pieces. NumPy incorporates vectorization extensively into its algorithms, offering concise, readable, and efficient code.

Broadcasting is equally essential. It defines how NumPy treats arrays with different shapes during arithmetic operations. Broadcasting enables vectorized array operations, reducing the need for explicit looping and indexing in Python. This results in more Pythonic, efficient, and concise code. Broadcasting is a concept that has its roots in mathematical procedures and has found its way into modern programming languages like Python and libraries like NumPy.

Calculus for Derivatives and Gradients

When discussing neural networks, one cannot evade the foundational role of calculus, particularly the concept of derivatives. In its simplest form, a derivative describes how a function changes as its input varies. Understanding derivatives is vital in neural networks because they provide insights into the direction and magnitude of changes we should apply to improve model performance.

To appreciate the significance of derivatives in neural networks, consider the analogy of a mountaineer seeking the lowest point in a hilly terrain. The slope informs each step the mountaineer takes of the terrain at that point. Similarly, a neural network adjusts its parameters in response to the slope, or gradient, of the loss surface, aiming to find the minimum loss.

A neural network comprises weights and biases that are continually adjusted during training to minimize the loss. This adjustment process, called gradient descent, leans heavily on derivatives. The gradient of the loss function concerning a particular weight gives the direction and magnitude of change required for that weight. If the gradient is positive, the weight should decrease to minimize the loss, and vice versa.

Gradients are vectors that point in the direction of the steepest ascent of a function. In a multi-dimensional space, such as the one inhabited by neural networks with many parameters, gradients become invaluable. Each gradient vector component corresponds to the loss function's derivative concerning a particular parameter. Following the negative gradient, we move towards the function's minimum—a foundational principle in training neural networks.

INTRODUCING GRADIENT DESCENT AND BACKPROPAGATION ALGORITHMS

Unveiling the heart of neural network optimization, we arrive at two indispensable algorithms: Gradient Descent and Backpropagation. Although sounding esoteric, these algorithms are the practical embodiments of the mathematical principles we've previously discussed.

While Gradient Descent provides a roadmap to adjust network parameters systematically to minimize loss, Backpropagation ensures the efficient computation of gradients across layers, making deep learning feasible. This section will dissect their workings, drawing connections between the foundational calculus and these algorithms' operational logic. Ready your intellectual arsenal as we dive deep into the mechanisms that empower neural networks to learn from data and improve over time.

Step-by-step Explanation of Gradient Descent

Gradient Descent is the backbone of numerous optimization problems in machine learning, and understanding its mechanics is paramount for anyone delving into neural networks. Let's systematically break down this pivotal algorithm.

1. **Objective:** At its core, Gradient Descent seeks to find the minimum of a function. In the context of neural networks, this function is the loss or cost function, which quantifies how far off our model's predictions are from the actual values.

2. **Initialization:** Begin with random values for the network's parameters (weights and biases). This can be

considered an arbitrary starting point on our function's surface.

3. Compute the Gradient: Evaluate the gradient of the loss function concerning each parameter. Recall that the gradient is a vector pointing toward the steepest ascent of our function. Each component of this gradient corresponds to the derivative concerning a particular parameter.

4. Update Rule: Use the computed gradient to adjust each parameter by a small step. The learning rate, a hyperparameter, governs the size of this step. If the gradient component is positive, the corresponding parameter decreases; if negative, it increases. The formula for the update can be expressed as:

"parameter (new) =parameter (old) −learning rate×gradient"

5. Iterate: Repeat the process of computing the gradient and updating the parameters for a set number of iterations or until the change in the loss function between iterations is negligibly small.

6. Learning Rate Nuances: The learning rate plays a critical role. A tremendous learning rate might overshoot the minimum, causing oscillations or divergence. Conversely, a minimal learning rate might result in excessively slow convergence. Fine-tuning this value is crucial for efficient optimization.

Backpropagation Algorithm for Updating Weights and Biases

Backpropagation, often termed *"backprop"*, is the linchpin algorithm that enables efficient weight and bias updates in neural networks. To comprehend backprop, imagine a cascading effect of errors flowing backward through the network, adjusting parameters.

1. Feedforward Pass: Initiate by feeding an input through the network, layer by layer, until an output is produced. This pass involves matrix multiplications and activation functions, resulting in a predicted output.

2. Compute the Loss: Calculate the difference between the predicted output and the actual target using a loss function. This value measures the network's error for that specific input.

3. Reverse Pass: Starting from the output layer and moving backward to the input layer, compute the loss gradient concerning each parameter. This is achieved by applying the chain rule of calculus, essentially breaking down the computation of complex derivatives into simpler, manageable parts.

4. Weight and Bias Update: With the computed gradients, apply the Gradient Descent update rule to adjust the weights and biases of each layer. The gradient and the learning rate determine the magnitude and direction of adjustment.

5. Iterative Refinement: Repeat this process for every input in the training dataset. As iterations progress, the weights and biases converge to values that minimize the network's overall loss.

EXPLORING WEIGHT AND BIAS UPDATES

Diving deeper into the neural network's inner workings, we come to a crucial juncture: the intricate art of updating weights and biases. These parameters, though minuscule in their individual capacity, collectively determine the prowess of the network. Their appropriate tuning is essential to metamorphose a naïve, untrained network into a predictive powerhouse.

Learning Rate and its Impact

The learning rate, often denoted by α or η, is a pivotal hyperparameter in neural network training. While the gradient provides direction, the learning rate dictates the magnitude of steps taken during optimization. The learning rate controls how much we adjust the model's weights and biases in response to the computed error. Simply put, it scales the gradients before the update. Mathematically, the update rule incorporates the learning rate as:

"parameter (new) =parameter (old) −learning rate×gradient"

A judiciously chosen learning rate ensures convergence towards the loss function's local or global minimum. Too large or too small, and the consequences can be dire. Recognizing the challenges of a static learning rate, many advanced optimization algorithms, like Adam, RMSprop, and AdaGrad, have been devised. These algorithms adjust the learning rate dynamically based on historical gradient information, ensuring smoother and often faster convergence.

High Learning Rate

Venturing into learning rates, one quickly recognizes the criticality of this parameter's magnitude. Opting for a high learning rate can expedite and hinder the model's training, presenting a paradox of outcomes. A notable advantage of a high learning rate is the potential for swift movement through the parameter space.

By making more significant adjustments to weights and biases, the model may quickly converge to a local minimum, reducing the overall training time. Overshooting is the primary peril of an elevated learning rate. Giant steps might leapfrog over minima, causing the model to oscillate back and forth without settling. Instead of a smooth descent, one witnesses a chaotic zigzag pattern, which might never stabilize.

Interestingly, a high learning rate can sometimes be beneficial by propelling the model out of plateau regions or shallow local minima, where gradients are nearly zero. In extreme cases, the loss might escalate with each iteration instead of converging, causing the model to diverge. This phenomenon results from excessively aggressive weight and bias updates. While a high learning rate promises speed, it comes with potential instability. The challenge lies in discerning the trade-off: leveraging the speed advantage without compromising the integrity and stability of the training process.

Low Learning Rate

Navigating the delicate balance of neural network training, a low learning rate emerges as a contrasting strategy to its high counterpart. While characterized by its cautious approach, the implications of such a choice are multifaceted and profound. A primary advantage of a low learning rate is the stability it introduces. More minor, incremental adjustments to weights and biases ensure a more controlled descent down the loss function's surface. The chance of overshooting minima or wild oscillations is substantially reduced.

The flip side of this stability is the pace of training. Small steps, although safe, can lead to protracted convergence times. This means more epochs might be required for the model to reach a satisfactory performance level. With conservative updates, the model might settle into shallow local minima or plateau regions without the momentum to escape. These points might not represent the optimal solution, leading to subpar model performance. A low learning rate can benefit specific applications that demand high precision.

It allows the model to fine-tune its parameters, inching closer to the exact minimum of the loss function. Due to the slow nature of convergence, training with a low learning rate can be resource-intensive, demanding more computational power and time. A low learning rate offers a meticulous approach to neural network training, ensuring stability and precision. However, this comes at the cost of extended training durations and potential entrapment in suboptimal solutions. Careful tuning and consideration are imperative to strike the right balance.

Finding the Right Learning Rate

Striking the ideal balance for the learning rate is equivalent to navigating a ship through turbulent waters. Too fast or too slow, and the journey becomes inefficient or unsafe. Here's a guide to optimizing this crucial hyperparameter:

Start by testing a range of learning rates, often on a logarithmic scale. For instance, you might evaluate rates at $10^1, 10^2, 10^3$, and so forth. Monitor the model's convergence and validation performance for each. Instead of a static rate, employ a dynamic approach where the learning rate decreases over time. Popular strategies include step decay, where the rate drops at specific epochs, and exponential decay, where it diminishes at a constant factor. Some modern frameworks offer tools to identify a suitable learning rate rapidly. By incrementally increasing the rate and plotting the loss, one can observe a sweet spot where the loss decreases rapidly before becoming unstable.

Combine a higher initial learning rate with early stopping mechanisms. If the model's performance deteriorates on a validation set, halt training. This strategy exploits the rapid convergence of high rates while curtailing potential divergence. Analyze the training loss curve. A smooth, descending curve suggests an apt learning rate, while oscillations or plateaus may hint at too high or too low rates, respectively. Determining the optimal learning rate isn't a one-size-fits-all endeavor. It demands experimentation, intuition, and patience, ensuring the chosen rate aligns with the model's architecture and the problem's intricacies.

Activation Functions and Their Derivatives

Within the neural network architecture, activation functions serve as gatekeepers, modulating and transforming the flow of information between neurons. They introduce non-linearity, enabling networks to capture complex patterns and relationships. But their role doesn't end there.

Sigmoid Activation Function

The sigmoid activation function is denoted mathematically as:

$$\text{``}\sigma(x) = 1/1 + e^\wedge - x\text{''}$$

It is a classic choice in neural networks. Mapping any input to a value between 0 and 1 is beneficial for binary classification tasks. Its characteristic S-shaped curve smoothly transitions from near-zero to near-one values.

However, it's not without caveats: the sigmoid function can cause vanishing gradient issues in deep networks, which may impede effective learning. Regardless, understanding its behavior and properties remains foundational in neural network theory.

ReLu (Rectified Linear Unit) Activation Function

The ReLU function is defined as

$$"f(x)=max(0,x)"$$

It has become a staple in modern neural network architectures. Favorable for its computational efficiency, ReLU introduces non-linearity without being bounded. This ensures positive activations pass through unaltered while negative ones are set to zero. While it mitigates the vanishing gradient problem, ReLU is not immune to issues—most notably, the "dying ReLU" phenomenon where neurons can sometimes become inactive during training and cease to update.

Tanh Activation Function

The hyperbolic tangent or tanh function, expressed as tanh(x), operates like the sigmoid but with an output range from -1 to 1. This centered range often proves beneficial, leading to faster convergence in training since outputs are zero-centered. Represented by an S-shaped curve, tanh offers non-linearity, making it suitable for various tasks. However, like the sigmoid, it can suffer from vanishing gradient problems, particularly when inputs lie in the extremes of its domain.

Other Activation Functions

Beyond the commonly employed sigmoid, ReLU, and tanh, the realm of neural networks boasts a plethora of activation functions, each tailored for specific challenges and scenarios. These include Leaky ReLU, which addresses the dying neuron issue of its predecessor, and Swish, which often outperforms traditional functions in deeper networks. Exponential Linear Unit (ELU) and Softmax, commonly used in multi-class classification, are other notable mentions. Exploring and understanding these diverse functions can unlock nuanced performance improvements in various neural architectures.

OVERVIEW OF COMMON LOSS FUNCTIONS AND OPTIMIZATION TECHNIQUES

The efficiency and efficacy of a neural network hinge upon judiciously chosen loss functions and optimization techniques. These components dictate how the network evaluates its errors and refines its parameters. Dive into this section to unravel the pivotal loss functions used across varied tasks and explore the optimization algorithms that drive convergence, ensuring neural models achieve their objectives precisely and quickly.

Mean squared error (MSE)

The Mean Squared Error (MSE) is a quintessential loss function predominantly used in regression tasks. It is mathematically expressed as:

$$MSE = \frac{1}{n} \sum_{i=1}^{n} \left(Y_i - \hat{Y}_i \right)^2.$$

where Y(i) represents the actual value, Ŷ(i) signifies the predicted value, and N is the total number of data points. The essence of

MSE is to determine the average squared differences between predicted and actual values. A notable characteristic of MSE is its emphasis on more significant errors. Squaring the discrepancies before averaging inherently gives more weight to substantial errors, compelling the model to address and correct these significant deviations more fervently.

While it's a popular choice for tasks like linear regression and time series forecasting, it's essential to note that its sensitivity to substantial errors can be both an advantage and a drawback. This sensitivity might render the model vulnerable to outliers, suggesting that alternative loss functions or data preprocessing techniques might be more suitable in situations with numerous outliers.

Cross-entropy loss

Cross-entropy loss, often associated with classification problems, quantifies the difference between two probability distributions: the actual distribution and the predicted one. It measures the average number of bits needed to identify an event from a set of possibilities. In the context of neural networks, minimizing cross-entropy implies pushing the predicted probabilities closer to the actual labels. This loss is particularly advantageous as it heavily penalizes confident yet incorrect predictions, thus guiding the model towards accurate classification.

It reduces to the binary cross-entropy for binary classification, whereas for multi-class tasks, the summation extends over all classes. It's crucial to understand that while cross-entropy is exceptionally effective for classification tasks, it assumes the model's outputs are probabilities, necessitating activation functions like softmax in the final layer.

Stochastic Gradient Descent (SGD) and its variants

Stochastic Gradient Descent (SGD) is an optimization technique that updates model weights using only a single data point or a small batch at each iteration rather than the entire dataset. This approach offers faster convergence, can escape local minima, and introduces update variance.

To address this, several variants of SGD have been developed, including Momentum, which accumulates velocity from previous gradients to smooth updates; Adagrad, which adjusts learning rates based on historical gradient values; and Adam, which combines elements of both Momentum and Adagrad, offering adaptive learning rates and smoother weight updates. These variants enhance SGD's efficacy, making it more robust and efficient.

In mastering the foundational elements of neural networks, one gains the knowledge necessary to transition from theory to implementation.

Having delved deep into the mathematical intricacies and foundational principles of neural networks, you are now well-prepared to undertake the hands-on journey of coding and implementation.

In the next chapter, we will transform abstract concepts into tangible Python code, bridging theory with practice and unlocking the true power of neural computation.

## 3

IMPLEMENTING NEURAL NETWORKS IN PYTHON

"Just because something doesn't do what you planned it to do doesn't mean it's useless."

— THOMAS EDISON

Neural networks stand as both a marvel and a challenge. In this chapter, we'll look at how to use these intricate architectures in Python. We will cover everything from setting up a conducive Python environment and building a basic feedforward network to tackling coding intricacies and robustly evaluating your models. Throughout this chapter, you'll find step-by-step processes and examples that'll help you implement neural networks in Python. So, let's get started.

SETTING UP THE PYTHON ENVIRONMENT FOR NEURAL NETWORK DEVELOPMENT

When mastering neural networks, a foundational step you cannot overlook is setting up a robust Python environment. This is important in establishing a stable, consistent, and isolated workspace, ensuring that every piece of code we run behaves predictably.

You need to understand that neural networks often require specific versions of libraries to function optimally, which makes relying on such factors critical. You see, having a chaotic or cluttered environment can lead to unforeseen errors, resulting in inconsistencies, and require additional troubleshooting time.

When you set up the Python environment correctly, it ensures that the functionality of all necessary dependencies and libraries is synchronized. Doing so allows you to minimize unexpected errors and increase productivity. In addition, you also need to know that utilizing tools like Anaconda can further streamline the process, handling package management and environment isolation.

This is necessary for you as a developer as it allows you to focus on the intricacies of your models. This section will dive into four key things instrumental in setting up an ideal Python environment for neural network development. This includes:

- **Installing Anaconda and Necessary Libraries:** Anaconda is a robust platform that simplifies package management, ensuring all dependencies are met and our libraries function cohesively.
- **Creating a Virtual Environment:** This allows for creating an isolated space where we can have specific versions of

packages without affecting the global Python setup. You need to know that such isolation is crucial, especially when working on multiple projects with differing requirements.

- **Activating the Virtual Environment:** Once created, it's vital to know how to activate our virtual environment, making it the active Python workspace for our current session.
- **Deactivating the Virtual Environment:** As vital as it is to activate, knowing how to deactivate and return to the global environment or switch to another virtual environment is equally crucial.

So, now that you know why setting the Python environment is crucial, let's dive deep into all the tasks mentioned above and see how you can do this on your own.

Anaconda, Necessary Libraries, And Virtual Environments

We'll learn how to install Anaconda and the necessary libraries in a second. But first, you need to know that this process is not as overwhelming as most people believe. Before we get into the exact steps you need to follow, you need to know that Anaconda is a free and open-source platform.

It lets you easily manage and utilize multiple data science and machine learning workflows. So, with that in mind, the steps you need to implement are as follows.

1. Download and Install Anaconda from the Official Website

The first thing you need to do is head over to Anaconda's official website. Once there, choosing the appropriate distribution for your operating system would be best. You can select a version for Windows, macOS, or Linux.

Once you've done that, ensure you're running the latest version of Python. For a seamless installation experience, just run the installer and follow the on-screen prompts.

2. Open Anaconda Navigator or Anaconda Prompt

Once the installation is complete, you can option the navigator by looking for the "Anaconda Prompt." You can also use the "Anaconda Prompt" for a command-line interface.

3. Create a Virtual Environment

The next step in the process is to create a virtual environment. Most people often overlook this aspect. However, it's crucial to understand how important this is. Creating a virtual environment helps you isolate different projects, allowing you to prevent package conflicts. To create a new virtual environment in the Anaconda Prompt, all you have to do is input the following command:

"conda create –name myenv"

4. Activate the Virtual Environment

Once you've created the virtual environment, you'll need to activate it. To do this, enter the following command in the Anaconda Prompt:

"conda activate myenv"

If you're using macOS or Linux, you can activate the virtual environment using the following command:

"source activate myenv"

It's important to understand that the prompt should change when you enter this command. This would reflect that you are now in the *"myenv"* virtual environment you've created.

5. Install Python Libraries

When installing Python libraries, there are two ways you can do this: using Conda or using Pip. Both of these are highly efficient at installing packages or modules for Python. However, you need to know that some libraries might be included in one manager and not the other.

If you want to proceed with installing the libraries using Conda, enter the following command:

"conda install library_name"

To use the Pip to install Python standard libraries, use the following command:

"pip install library_name"

Before you begin, you need to know that you don't have to spend much time identifying which Python libraries you need. That's what this book is for. Some of the most commonly used libraries for data science and machine learning workflows include:

- **NumPy** - this is a fundamental package that is essential for scientific computing in Python.
- **Pandas** - this library provides high-level data structures and methods for manipulation.
- **Matplotlib** - this library is essential as it enables data visualization.
- **Scikit-learn** - this is an essential machine-learning library that's packed with different algorithms and tools.

When using Conda, you can install all these libraries at once using the following command:

"conda install numpy pandas matplotlib scikit-learn"

You can start coding in Python using these libraries once installed. Creating a virtual environment, as mentioned above, will allow you to have the flexible Python environment needed to develop a neural network. Once the libraries have been installed, you must open the editor, import the libraries, and begin coding.

If you're using Windows and want to deactivate the virtual environment, you can do that by entering the following command:

"conda deactivate"

If you're on macOS or Linux, you can deactivate the virtual environment by using the following command:

"source deactivate"

Now that you know how to install libraries and create a virtual environment, let's look at how to create a basic feed-forward neural network using Python.

IMPLEMENTATION OF A BASIC FEEDFORWARD NEURAL NETWORK

Neural networks have, without a doubt, revolutionized machine learning and artificial intelligence. Before we dive into the foundational steps you need to take to construct and train a basic feedforward neural network, there are a few things you need to know.

Defining the architecture and key parameters is essential to the training. Forward propagation is where we input data through the network, producing an output. Backpropagation is the network's self-reflection phase, which better adjusts to capture underlying data patterns. So, with that in mind, let's look at each of these things in more detail.

Defining the Architecture and Parameters

Before we jump into coding, it's crucial to understand and define the architecture and parameters of our neural network. You must know that the architecture determines how data will move through the network. Parameters such as weights and biases are what you optimize during the training phase.

A basic feedforward neural network consists of an input layer, one or more hidden layers, and an output layer. These layers are connected using nodes, also referred to as neurons. Let's look at each layer in more detail.

- **Input Layer** - this layer receives the input features. The number of neurons in the input layer corresponds to the number of input features.
- **Hidden Layers** - these are positioned between the input and output layers. They help the network learn complex patterns. Each hidden layer can have a varying number of neurons, and choosing the optimal count often requires experimentation.
- **Output Layer** - outputs the final prediction or classification. The number of neurons in the output layer typically corresponds to the number of classes for classification problems or just one neuron for regression problems.

Each neuron processes the input data and transfers it to the next layer using an activation function. Some of the most essential activation functions you can use include:

- **Sigmoid** - this function limits the output between 0 and 1 and is suitable for binary classification.
- **ReLU (Rectified Linear Unit)** - this function outputs the input directly; if it is positive or outputs zero, it's not.

- **Softmax** - this function is used in the output layer of multi-class classification problems and allows you to give a probability distribution over classes.

You need to understand that every connection between neurons has an associated weight. These weights are adjusted during training to reduce the difference between the predicted and actual outputs. Biases, on the other hand, are like intercepts in linear regression. You need to know that they are used to ensure the neurons can still be activated even when input features are zero.

Another that you need to know is that hyperparameters are set before training starts and are not updated during training. Some of the most essential hyperparameters that you can use in a neural network include:

- **Learning rate** - this dictates the step size during optimization. You need to know that a high rate can cause the model to converge quickly, while a low rate may slow the process.
- **Batch size** is the number of samples processed before updating the model.
- **Number of epochs** - the number of times the learning algorithm will work through the entire training dataset.

You can use the NumPy library for matrix operations. You need to know that these operations are the core of the feedforward and backpropagation processes. Once you've imported the library, you must define each layer's number, weights, biases, and activation functions.

After that, you can create a function that calculates the output for the input you provide using the network's predefined weights and biases, as done in the example code below.

```
"import numpy as np
def sigmoid(x):
return 1 / (1 + np.exp(-x))
def feedforward(inputs, weights, biases):
"""
Calculate the output of a basic neural network with one hidden layer.
Parameters:
- inputs (list or ndarray): Input data.
- weights (list of ndarrays): Weights for each layer.
- biases (list of ndarrays): Biases for each neuron in each layer.
Returns:
- output (ndarray): Output of the network.
"""
# Input to Hidden Layer
hidden_layer_input = np.dot(inputs, weights[0]) + biases[0]
hidden_layer_output = sigmoid(hidden_layer_input)
# Hidden Layer to Output
output_layer_input = np.dot(hidden_layer_output, weights[1]) + biases[1]
output = sigmoid(output_layer_input)
return output"
```

Forward Propagation And Computing Activations

Forward propagation is the initial phase in the training of a neural network. In this phase, the input data goes through the network one layer at a time until it produces an output. This process involves the computation of the activations for each neuron in a prediction at the output layer.

The difference between this prediction and the actual target is what backpropagation seeks to minimize by adjusting weights and biases. Before we get into the coding bits, you need to understand that the forward propagation process includes three key factors: inputs, weighted sum, and the activation function.

It's important to know that forward propagation provides an initial prediction based on current weights and biases. This sets the stage for backpropagation, which uses the error from the forward propagation's output to adjust the weights and biases. So, with that in mind, let's look at the factors we've mentioned above.

Inputs - this process begins with the feature data, which is the initial input.

Weighted Sum - Calculate a weighted sum of each neuron's inputs. This is achieved by taking the dot product of the input data with the neuron's weights and adding the bias.

"Weighted Sum = (input \times weights) + bias"

Activation Function - The weighted sum is then passed through an activation function to compute the neuron's activation (output). Common activation functions include the sigmoid, ReLU, and softmax.

"Activation = f(Weighted Sum)"

Propagation - this activation becomes the input for the next layer's neurons, and the process repeats until the final layer produces the network's output. So, with that in mind, let's look at how this would look in Python.

```
"import numpy as np
def sigmoid(x):
return 1 / (1 + np.exp(-x))
def forward_propagation(inputs, weights, biases, activation_funcs):
"""
Computes activations for each layer using forward propagation.
Parameters:
- inputs (ndarray): Input data.
- weights (list of ndarrays): Weights for each layer.
- biases (list of ndarrays): Biases for each neuron in each layer.
- activation_funcs (list of functions): Activation functions for each layer.
Returns:
- activations (list of ndarrays): Activations for each layer.
"""
activations = [inputs]
for w, b, func in zip(weights, biases, activation_funcs):
z = np.dot(activations[-1], w) + b
a = func(z)
activations.append(a)
return activations"
```

Backpropagation And Weight Updates

Backpropagation is the backbone of training a neural network. While forward propagation computes the activations and outputs, backpropagation focuses on minimizing the error by adjusting the weights and biases of the network. The core idea is to update the weights in the direction that reduces the total error. The key thing you need to understand about backpropagation include:

- **Loss Calculation** - After obtaining the output from forward propagation, we need to calculate the loss (or error). The choice of loss function depends on the nature of the task. So, for example, you could use mean squared error for regression and cross-entropy for classification.

- **Gradient Computation** - The crux of backpropagation lies in computing the loss gradient concerning the weights and biases. The gradient signifies how the loss would change if a slight deviation in the weights or biases occurs.
- **Chain Rule** - To compute these gradients, backpropagation leverages the chain rule from calculus. It breaks down the overall loss function into simpler parts and computes the gradient layer by layer, moving backward from the output to the input.

To master neural network training, you must understand how gradients guide adjustments in network parameters. Understanding the gradient's role at each layer lets you tweak the network to fit your data effectively. Let's examine the gradient calculations across different layers and their impact on weights and biases.

- **Output Layer Gradient** - For the output layer, we first calculate the derivative of the loss function concerning the output activation. This result is then multiplied by the derivative of the activation function pertaining to its input.
- **Hidden Layers Gradient** - For hidden layers, the gradient is computed based on the gradient from the subsequent layer and the derivative of the activation function. It's a recursive process that moves from the last hidden layer towards the input layer.
- **Weight & Bias Update** - Once gradients are computed for all layers, the weights and biases are updated using an

optimization algorithm like Gradient Descent. The general update rule is:

"new_weight = old_weight - learning_rate X gradient"

Now that you understand what backpropagation is all about and know the role gradients have in each layer let's look at how this would play out in Python using a sample code.

```
"import numpy as np
def sigmoid(x):
return 1 / (1 + np.exp(-x))
def sigmoid_derivative(x):
return sigmoid(x) * (1 - sigmoid(x))
def backpropagation(inputs, outputs, weights, biases, learning_rate=0.01):
"""
Performs a backpropagation step to update weights and biases.
Assumes a simple network with sigmoid activations and MSE loss.
"""
# Forward pass
z_hidden = np.dot(inputs, weights[0]) + biases[0]
a_hidden = sigmoid(z_hidden)
z_out = np.dot(a_hidden, weights[1]) + biases[1]
a_out = sigmoid(z_out)
# Compute error at the output
error_out = (a_out - outputs) * sigmoid_derivative(z_out)
# Compute error at the hidden layer
error_hidden = error_out.dot(weights[1].T) * sigmoid_deriva-
tive(z_hidden)
# Update weights and biases
weights[1] -= learning_rate * a_hidden[:, None].dot(error_out[None, :])
biases[1] -= learning_rate * error_out
```

*weights[0] -= learning_rate * inputs[:, None].dot(error_hidden[None, :])*
*biases[0] -= learning_rate * error_hidden*
return weights, biases"

ADDRESSING COMMON CODING CHALLENGES

Embarking on implementing neural networks, especially for beginners, is similar to navigating uncharted waters. The complexity of these architectures and the intricacies of data handling can lead to some challenges. Some of them might include

- Misaligned data shapes cause dimensionality errors.
- Unintentional architectural configurations lead to inefficient models.
- Overfitting is where a seemingly perfect model performs poorly on new data.

Debugging Techniques and Error Handling

The challenges beginners face while building a neural network can limit their achievement. However, you must understand that these challenges can be addressed using different debugging and error-handling techniques. Let's take a detailed look at what they are.

Inspect Data and Preprocessing

Before diving into the intricacies of model training, a keen inspection of data is paramount. Many issues in neural network performance stem from inconsistent or incorrectly preprocessed data. It would be best to begin by scrutinizing the initial few rows of your dataset and examining them for missing or abnormal values.

You can visualize distributions to spot outliers or skewed variables. If you've used any, ensure that data normalization or standardization processes are consistent across training and testing sets. Taking the time to validate your preprocessing pipeline can prevent countless headaches down the line.

Verify Model Architecture

Your neural network's architecture is pretty similar to the foundation of a building. What this means is that any misalignment could lead to catastrophic results. Therefore, you need to verify that the input layer matches the shape of your processed data.

Ensure activation functions align with your task, especially in the output layer. For example, using softmax for multi-class classification and sigmoid for binary classification. In addition, you need to ensure that the loss function complements your task and activation functions.

Print Model Summary

Most deep learning frameworks provide a mechanism to print out a concise summary of your model. This summary reveals layer-by-layer architecture, showcasing the number of trainable parameters, output shapes, and more.

By reviewing this, you can identify unintended redundancies, misconfigurations, or potential bottlenecks. It offers a quick and holistic view, ensuring the model's structure aligns with your intentions.

Debug Training Process

Observing the model's training behavior can give you many insights. If the loss doesn't decrease or fluctuates significantly, there might be issues with the learning rate, initialization of weights, or data inconsistencies.

Pay attention to metrics during training, and be wary of situations where training loss decreases while validation loss increases – this could signal overfitting. Always maintain a structured logging mechanism to trace and analyze the training evolution.

Handle Overfitting and Underfitting

Overfitting occurs when a model performs exceptionally on training data but poorly on unseen data, while underfitting occurs when the model doesn't capture the underlying pattern of the data.

Techniques like dropout, early stopping, or L1/L2 regularization can mitigate overfitting. On the other hand, if a model under fits, consider making the architecture more complex or re-examining data preprocessing techniques. Additionally, cross-validation can help identify and handle these challenges early on.

Implement Error Handling

Unanticipated errors are common in the code-heavy environment of neural networks. Implementing error handling techniques, especially using *"try"* and *"except"* blocks in Python, can guide you to the root cause of the issue. This helps catch runtime errors and provides insightful error messages, making debugging more efficient and less frustrating.

Use Debugging Tools

Modern IDEs and programming environments come equipped with robust debugging tools. The use of breakpoints, allowing you to pause execution and inspect variables' states, can be invaluable.

Variable inspection provides a snapshot of current data, helping to pinpoint where values might be going astray. Leveraging these tools can dramatically reduce the time and effort spent identifying and rectifying issues.

Visualize and Analyze Results

Once your neural network is trained, don't rely on numerical metrics. Visualization tools can provide profound insights, like plotting the confusion matrix, ROC curve, or even specific layers' activations and feature maps.

They help identify which classes or features the model struggles with, guiding further iterations and refinements. Always remember that visual insight can often convey what raw numbers cannot.

Efficient Coding Practices for Performance Optimization

The world of neural networks is not just about designing intricate architectures or refining hyperparameters. The engine of efficient code is at the heart of any successful deep-learning project.

Writing optimized code is similar to fine-tuning cars; it ensures the fastest speeds and optimal performance, delivering results in record time. So, with that picture in mind, let's deep dive into some stellar coding practices that can significantly elevate the performance of your neural networks.

- **Avoid Global Variables**

You need to know that global variables are easily accessible but can slow down performance as they are stored in a global memory space. Instead, use local variables within functions, as they reside in the stack and provide faster access.

- **Use Built-in Functions and Libraries**

Languages like Python come with built-in functions and libraries optimized for performance. For instance, leverage Python's built-in "sorted()" function rather than writing a sorting algorithm from scratch.

- **Limit Memory Consumption**

Inefficient memory usage can cause algorithms to run slower. Always free up memory by deleting objects or variables no longer in use. Remember, the "del" statement can be a handy tool in Python.

- **Vectorize Your Computations**

Instead of using loops, leverage vectorized operations provided by libraries like NumPy. You need to know that these libraries are typically implemented in C or Fortran behind the scenes, offering much faster execution than native Python loops.

- **Profiling Code**

Understand where your code spends the most time using tools like Python's *"cProfile."* Profiling code is critical because you can optimize those areas once you know the bottlenecks.

- **Efficient Data Structures**

Selecting the proper data structure, be it a list, set, dictionary, or another can dramatically impact performance. Think about it: checking membership in a set is faster than in a list.

- **Use JIT Compilers**

Another thing you need to know is that Just-In-Time (JIT) compilers, like Numba, can substantially speed up Python code, especially when dealing with loops or functions with heavy computations.

- **Avoiding Recursion**

Recursion can make code look clean, but it might not always be the most efficient approach, especially for deep recursions. Iterative solutions can often be faster and consume less memory.

- **Leverage Efficient Algorithms**

Remember the basics! An efficient algorithm can significantly reduce computational complexity. For instance, using quick sort over bubble sort can make a difference in execution times.

- **Parallel Processing**

For tasks that can be split, you need to use parallel processing to distribute the load and execute multiple tasks simultaneously. This is especially relevant for deep learning models, which can be trained on various GPUs.

With these practices in hand, not only will your neural network implementations run smoother, but you'll also save valuable computational resources and time. Now that you comprehensively understand how to implement neural networks in Python let's look at how you can test and evaluate them.

TESTING AND EVALUATING NEURAL NETWORK MODELS

After learning about setting up your Python environment, building neural network architectures, managing potential coding challenges, and optimizing for performance, it's time we look at testing and evaluating your models.

You must understand that neural networks must be critically evaluated to ensure they perform as intended in real-world scenarios. The fundamental step in this process is splitting data into training and testing sets. Let's dive into this practice in detail.

Splitting Data Into Training And Testing Sets

In machine learning and deep learning, data is the most prized possession. It powers our models, provides insights, and ultimately determines the success or failure of our neural network. However, using the entire dataset to train our model can lead to a problematic scenario.

Without a separate dataset to test our model, we can't ensure that it generalizes well to unseen data and not merely memorizes the training data. This is where splitting your data comes into play. We divide our data into two distinct sets.

- Training Set - this is the primary dataset used to train the model. It helps the neural network learn the underlying patterns and relationships within the data.
- Testing Set - this dataset is kept separate and is not shown to the model during training. It's employed post-training to evaluate the model's performance on unseen data. This aids in assessing how well the model will likely perform in real-world scenarios.

To split the data, you need to import the necessary libraries. The "train_test_split" function from the "sklearn.model_selection" module is commonly used in Python. To use this, enter the following command:

"from sklearn.model_selection import train_test_split"

Once you've done that, apply the following function.

"X_train, X_test, y_train, y_test = train_test_split(X, y, test_size=0.2, random_state=42"

"X" and *"y"* are your features and target variables. The "test_size" parameter dictates the proportion of the dataset you wish to allocate to the testing set, which would be 20%. The *"random_state"* ensures reproducibility.

Once split, you can proceed to train your model on the *"X_train"* and *"y_train"* datasets and subsequently evaluate its performance using the *"X_test"* and *"y_test"* datasets.

Correctly splitting your data is a genuine measure of your model's performance. It ensures your evaluations' integrity, ensuring your neural network operates precisely and reliably.

Computing Accuracy And Other Evaluation Metrics

Accuracy isn't the sole determinant of a model's performance, especially in cases where data is imbalanced or false positives and negatives have different implications. Evaluating neural networks necessitates a holistic approach, incorporating various metrics to garner a comprehensive understanding of the model's prowess.

Accuracy provides a straightforward measure, calculating the ratio of correctly predicted instances to the total instances. However, it might be misleading in imbalanced datasets. To tackle such nuances, other metrics that can be used include:

- Precision - measuring the accuracy of optimistic predictions.
- Recall - highlighting the fraction of positives that were correctly identified.
- F1-Score - a harmonic mean of precision and recall.

For regression tasks, Mean Absolute Error, Mean Squared Error, and R-Squared offer insights into the model's deviation from actual values and its overall explanatory power.

Viewing a neural network's performance through multiple metrics, each offering unique insights, ensuring both robustness and reliability in real-world applications, is imperative.

And that's all to learn about implementing a neural network in Python. You start using the correct libraries and creating a virtual environment, and then on, it's to forward and backpropagation. In the coming chapter, we'll focus on understanding how to handle complex concepts in neural networks.

4

HANDLING COMPLEX CONCEPTS
IN NEURAL NETWORKS

"Just because something doesn't do what you planned it to do doesn't mean it's useless."

— THOMAS EDISON

Neural networks, a term once reserved for academia and specialized industries, have now become essential to modern technological advancements. In this chapter, we'll dive deep into the advanced neural network architectures, challenges that may exist when training models, and how you can use regularization techniques to address them.

We'll cover everything from the image-centric Convolutional Neural Networks to the time-sequential prowess of Recurrent Neural Networks and the might of deep learning models.

However, understanding these models is only half the battle. The key is to harness power without faltering into pitfalls like overfitting or underfitting.

EXPLAINING ADVANCED NEURAL NETWORK ARCHITECTURES

When you dig deeper into AI, you'll learn that advanced neural networks are essential for effective functioning. However, it's important to understand that each type of advanced neural network architecture has its own functionality and caters to different challenges. This section will review these architectures and learn more about their capabilities and limitations.

We'll start with Convolutional Neural Networks tailored for image processing tasks and then shift our attention to the Recurrent Neural Networks (RNNs) used for sequential data. Lastly, we'll focus on DL models adept at handling intricate and diverse tasks. A robust comprehension of the architecture will give you a solid foundation. This will be of benefit when you're practically implementing these architectures in various real-life scenarios.

Convolutional Neural Networks For Image Processing

A convolutional neural network (CNN) is one of the most important aspects of computer vision and image processing. Originating from a rich history of neural network evolution, CNNs have transcended traditional architectures due to their unparalleled efficiency in extracting hierarchical features from raw image data.

Before understanding how CNNs can be used for image processing, you must know that a CNN comprises three layers. The purpose of each layer varies from the other and is essential for efficient functioning. Let's look at each one in more detail.

1. **Convolutional Layer** - this layer performs the convolution operation, extracting features from the input image. The convolution operation involves applying a filter or kernel over the input data to produce a feature map or convolved feature.

2. **Pooling or Subsampling Layer** - once the convolution is complete, this layer helps reduce the spatial dimensions of the data. This helps reduce the computational complexity but also assists in detecting specific features invariant to scale and orientation changes.

3. **Fully Connected Layer** - after convolutional and pooling layers, the architecture often has one or more fully connected layers where neurons connect to all activations in the previous layer. This is similar to traditional multi-layer perceptrons and helps classify the images based on the high-level features learned by the convolutional layers.

The primary advantage of a CNN is its ability to learn spatial hierarchies automatically and adaptively. You must know that the initial layers might capture basic structures like edges and textures. However, it's the deeper layers that capture complex patterns and objects.

Such hierarchical extraction capabilities make CNNs perfect for image processing and recognition tasks. CNNs have emerged as the go-to architecture for various functions in image processing. Some of these tasks include:

- **Image Classification** - whether differentiating a cat from a dog or identifying thousands of object types, CNNs excel in categorizing images into predefined classes based on learned features.

- **Object Detection** - beyond just classifying an entire image, CNNs can localize and identify multiple objects within an image, providing both the object's label and its bounding box.
- **Image Segmentation** - CNNs can divide an image into multiple segments, each corresponding to an object or a part of an object. This makes it especially useful in tasks like medical imaging, where precise boundaries are required.
- **Facial Recognition** - by capturing faces' intricate features and patterns, CNNs can differentiate and recognize individual faces, even in varied lighting and poses.
- **Image Generation** - with architectures like Generative Adversarial Networks (GANs), which have CNN components, generating new, synthetic images that resemble a given dataset is possible.

While CNNs have proven instrumental in pushing the boundaries of image processing, they come with their challenges. Before we proceed further, it's essential for you to fully comprehend what these challenges are so that you may be able to address them when required.

- **Data Augmentation** - given the data-hungry nature of CNNs, the available image datasets are sometimes insufficient. Techniques like rotation, zooming, and flipping are employed to increase the dataset size artificially. However, you need to know that these can sometimes introduce noise.
- **Varying Image Resolutions** - cNNs can be sensitive to input image resolutions. Processing high-resolution

images can be computationally intensive. In addition, downsampling might result in the loss of critical features.

- **Invariance and Equivariance** - while CNNs are designed to handle spatial hierarchies and invariances, there are cases (like rotations or tilts) where additional techniques or data augmentation might be required to ensure consistent performance.
- **Overfitting on Image Data** - especially in cases where the number of parameters in the CNN far outweighs the number of training samples, there's a risk of overfitting, wherein the CNN performs exceptionally well on the training data but poorly on new, unseen images.

Recurrent Neural Networks (RNNs) For Sequential Data

Recurrent Neural Networks (RNNs) are pivotal in handling sequential data. The functionality of an RNN handling sequential data differs from that of a feedforward neural network. In a traditional feedforward neural network that assumes independence between inputs.

However, you need to know that RNNs possess an inherent memory mechanism. This mechanism allows them to remember and utilize past information in current computations. This characteristic suits them particularly for sequences where temporality and order are paramount.

An RNN processes sequences by iterating through the sequence elements and maintaining a state containing information relative to what it has seen. For every input in the sequence, the RNN first computes the output. Once that is complete, it updates its state, which will be passed to the next step of the sequence.

It's essential to comprehend that this state functions as the *"memory"* of the RNNs, meaning that is how RNN gets its recurrent nature. This memory-oriented nature of the RNNs ensures they can be used in tasks where sequence and order are of utmost importance. Some commonly known examples of these tasks include:

- **Time Series Forecasting** - whether it's predicting stock prices, weather patterns, or energy consumption, RNNs can model time dependencies to forecast future values based on past trends.
- **Natural Language Processing (NLP)** - texts are inherently sequential. RNNs excel in tasks like sentiment analysis, machine translation, and text generation, where the meaning of a word often depends on its preceding words.
- **Speech Recognition** - converting spoken language into text requires understanding audio sequences, and RNNs are the backbone for many state-of-the-art speech recognition systems.
- **Music Generation** - creating coherent and melodious tunes can be seen as a sequence generation task, and RNNs have been employed to produce novel musical compositions.

Using RNNs for these tasks and others like them does help individuals streamline specific processes and improve productivity. However, it's essential to acknowledge that this comes with certain limitations and challenges. For instance, working with RNNs can lead to vanishing and exploding gradient problems.

During training, RNNs often suffer from gradients that either vanish or explode, making the training process slow and sometimes unfeasible. In addition, standard RNNs struggle with long sequences. They tend to forget information from earlier time steps, making them less effective for tasks that require understanding over extended sequences.

However, problems triggered by a standard RNN short-term memory can be addressed using Long Short-Term Memory (LSTM) networks. LSTMs are a specialized kind of RNN that is equipped with memory cells and gates. These capabilities allow the LSTM to preserve information for longer sequences and discard it if it's irrelevant.

In addition, they efficiently update their state based on new information. Because of these capabilities, LSTMs and their variants, like GRUs (Gated Recurrent Units), have primarily overshadowed standard RNNs in many applications involving longer sequences.

Deep Learning Models For Complex Tasks

In the vast realm of artificial intelligence, complexity often alludes to tasks that aren't merely about linear separations or basic pattern recognition. They encompass intricate interdependencies and high dimensionality and often require a holistic understanding of data in various forms, such as images, texts, or sounds.

Deep learning uses neural networks with multiple layers that transform the data in successive stages, capturing increasingly abstract and detailed features. The depth of these networks allows for intricate pattern recognition, making them particularly potent for complex tasks. Deep learning is a sub-field of machine learning, but the two are drastically different, and understanding these differences is paramount.

ML models often work well with smaller datasets. They can produce good results with a few thousand data points. Meanwhile, DL models require vast amounts of data to function optimally. They typically thrive on big data, needing millions of data points to discern patterns effectively.

It's important to know that The complexity and depth of DL models require more data to generalize well, whereas traditional ML models can work with less data and fewer parameters. Generally, conventional ML models can run on standard CPUs and don't necessitate specialized hardware.

However, DL models, especially large neural networks, greatly benefit from GPUs or TPUs for training due to their parallel processing capabilities. The architecture of neural networks in DL, with its multiple layers and numerous nodes, demands high computational power, unlike many ML algorithms.

Many ML models, such as decision trees or linear regression, are inherently interpretable, making it easy to understand the relationships and decisions within the model. But DL models, on the other hand, due to their complexity, are often seen as *"black boxes,"* making it challenging to discern how they reach specific decisions. In addition, it's also vital for you to remember that:

- The simplicity of many ML models allows for more precise insights into their decision-making processes, whereas the intricate layers of DL models obscure their internal workings.
- The depth and breadth of neural networks and their data-hungry nature necessitate longer training durations.

- Deep learning's ability to self-learn representations from raw data sets it apart from traditional ML, where manual intervention is often needed to define data features.
- The essence of deep learning lies in its depth, using multiple layers to understand hierarchies and patterns in data, while ML can often capture patterns with more superficial structures.

DL is the fundamental building block of different neural networks for complex tasks. These neural networks included CNNs and RNNs, which we have discussed previously. However, deep learning can also be used for Generative Adversarial Networks (GANs) and Transformer Architecture.

GANs are innovative architecture that pits two networks, a generator and a discriminator, against each other. GANs shine in complex tasks like image generation, style transfer, and even generating art. Transformer architectures, on the other hand, are used in NLP.

Transformers and their progenies, like BERT, GPT, and more, tackle intricate challenges ranging from question-answering systems to achieving human-like performance in games. DL models can be used in various tasks, some mentioned below.

- **Medical Diagnoses** - Deep learning models analyze medical images, detecting tumors, anomalies, and diseases with a precision that sometimes surpasses human experts.
- **Voice Assistants & Translation** - Deep learning lies at the heart of today's advanced voice assistants, from understanding diverse accents to translating languages in real time.

- **Augmented Reality (AR) and Virtual Reality (VR)** - Deep learning aids in object recognition and environment mapping, ensuring realistic and interactive AR/VR experiences.
- **Autonomous Vehicles** - From detecting pedestrians to making split-second decisions, deep learning algorithms ensure safety and efficiency in self-driving cars.

However, like with anything else, using DL for complex tasks comes with unique challenges. For instance, DL models, especially the intricate ones, demand substantial computational resources in terms of memory and processing power. The sophisticated nature of these models often requires vast amounts of labeled data for training.

Furthermore, DL models are more complex than ML models, making understanding their decision-making processes challenging. However, these challenges can be addressed with the evolution of DL frameworks like TensorFlow and PyTorch and advancements in hardware accelerators.

INTRODUCING REGULARIZATION TECHNIQUES

While it's tempting to design neural networks that perform exceptionally well on training datasets, the actual test of a model's prowess lies in its ability to generalize to new, unseen data. You need to know if a model becomes too tailored to a training dataset; variations in the data may lead to overfitting.

However, there are regularization techniques you can use to prevent models from being overly tailored to the training set and ensure their robustness in real-world scenarios. In this section, we'll look at two crucial regularization strategies that have shaped the foundations of practical neural network training.

Dropout Regularization To Prevent Overfitting

Maintaining a balance between the model's learning capability and its generalization is paramount in the dynamic landscape of neural networks. It's important to know that as you encounter a model's accuracy, you'll likely experience an overfitting problem. However, it would be best to remember that overfitting can be addressed through dropout regularization.

This method randomly sets a fraction of input units to 0 at each training update. So it's essentially a neural network that, during training, drops out random neurons, ensuring that no single neuron becomes overly specialized or reliant on its neighboring neurons.

Conversely, overfitting happens when a neural network or any model learns the training data too well, including its noise and outliers. Most individuals commonly mistake this for an advantage. It may perform well on the training data. However, it's crucial to comprehend that the model or network will struggle with new data.

The underlying cause for such a scenario is that the model is that the training tailors the model to the data sets provided during that phase. Any difference within the new data then negatively impacts the model's performance.

Dropout offers a unique solution to this dilemma. Dropouts force the network to distribute information across various paths by randomly deactivating specific neurons during training. It ensures that the model doesn't become too dependent on any particular neuron, promoting redundancy in learning. In other words, training multiple neural networks and then averaging their predictions during inference makes the model more robust.

Implementing Dropout Regularization

By now, you have an understanding of what dropout regularization and overfitting are and how the two are interlinked with each other. With this in mind, let's look at the steps to utilize dropout regularization to counter overfitting.

1. The first thing you need to do is choose a dropout rate. This rate represents the fraction of neurons to be dropped out. Commonly used values range from 0.2 to 0.5. For example, if you have a layer with 100 neurons and choose a dropout rate of 0.3, approximately 30 neurons will be randomly dropped during each training iteration.

2. Once you've done that, apply the dropout operation to one or more layers in the neural network. It's typical to apply dropout after the activation function of a layer.

3. Now, during training, at each iteration, randomly deactivate specific neurons based on the dropout rate.

4. During evaluation or inference, do not apply dropout when you are using the model for prediction on unseen data. At this stage, all neurons should be active and contribute to predictions.

If you were to use TensorFlow and Keras and implement this in Python, this is what the syntax would look like:

```
"from keras.models import Sequential
from keras.layers import Dense, Dropout
model = Sequential()
# Adding a dense layer
model.add(Dense(128, activation='relu', input_shape=(input_di-
mension,)))
# Applying dropout regularization
model.add(Dropout(0.3))
# Continue with the rest of the model
model.add(Dense(64, activation='relu'))
model.add(Dense(output_dimension, activation='sigmoid'))"
```

However, before using dropout regularization, you need to consider the adaptive learning rate, regularization strength, and the position of the dropout layers. Dropout can affect convergence. So, it would be best to consider using adaptive learning rate algorithms like Adam when training with dropouts.

In addition, you can also experiment with the dropout rate. A higher rate means more regularization but can slow down training. Regarding the position, dropout can be applied to most layers. It's commonly used after fully connected layers. However, its positioning should be determined based on the problem at hand.

L1 AND L2 REGULARIZATION FOR WEIGHT DECAY

Diving deeper into the domain of regularization techniques, two pivotal methods have garnered substantial attention in machine learning: L1 and L2 regularization. These techniques are crucial in mitigating the effects of weight decay and ensuring robust model training.

L1 Regularization and Lasso Regularization add a penalty equivalent to the absolute value of the magnitude of the coefficients. This type of regularization can lead to zero coefficients, meaning that the model entirely ignores some of the features. L2 Regularization, Ridge Regularization, adds a penalty equivalent to the square of the magnitude of coefficients.

The same factor shrinks all coefficients. However, none of them are eliminated. When represented mathematically, this would look like:

- L1:Loss=OriginalLoss+$\lambda\sum$|weights|
- L2:Loss=OriginalLoss+$\lambda\sum$weights2

In the equations above, lambda would be the strength of the regularization. Before proceeding, you need to know that weight decay predominantly refers to adding an L2 penalty to the loss function. Adding this penalty encourages the model to keep the weights small. Weight decay is a regularization method that prevents weights from growing too large and causing overfitting. It does so by adding a penalty proportional to the size of the weights.

Both L1 and L2 regularization techniques aim to penalize specific configurations of weights, ensuring the model does not fit too closely to the training data, making it more generalized and less susceptible to noise. Meanwhile, L2 regularization penalizes the square value of weights, encouraging smaller weights but not necessarily zero.

L1, on the other hand, can push some weights to become exactly zero, effectively performing feature selection. You can implement the L1 and L2 regularization using the steps mentioned below:

1. Decide which type of regularization (L1, L2, or both) is appropriate for your problem.
2. Choose a value for lambda, the regularization strength—a more considerable lambda results in more significant regularization and smaller weights.
3. Add the regularization term to the loss function and adjust the model training process to account for this added term.

If you implement L1 and L2 regularization using TensorFlow and Keras in Python, the syntax will look like this:

"from keras.models import Sequential
from keras.layers import Dense
from keras.regularizers import l1, l2
model = Sequential()
 # Adding a dense layer with L1 regularization
model.add(Dense(128, activation='relu', kernel_regularizer=l1(0.01),
input_shape=(input_dimension,)))
 # Adding another dense layer with L2 regularization
model.add(Dense(64, activation='relu', kernel_regularizer=l2(0.01)))
 # Finish the model architecture
model.add(Dense(output_dimension, activation='sigmoid'))"

However, before implementing these regularization techniques, you should consider different factors, including the scale of data, lambda, and the combination of regularization. It's essential to scale input data before applying regularization. You need to know that regularized algorithms are sensitive to the scale of input data.

The choice of lambda is crucial since a value too large might underfit the data, while a value too small might not regularize effectively. In addition, L1 and L2 can be combined using Elastic Net Regularization, which might be beneficial in scenarios where features are correlated.

TACKLING OVERFITTING, UNDERFITTING, AND MODEL CAPACITY

Tackling overfitting and underfitting is crucial in neural network training. Overfitting indicates a model too aligned with training data, compromising generalization. Underfitting shows a model's failure to capture data nuances. However, regularization techniques can be used to address these problems.

Early Stopping And Model Selection

Within the extensive world of neural networks, an often overlooked yet profoundly impactful to consider is when we should cease the training of our model. Though seemingly beneficial, continuous training can lead to overfitting, where the model becomes excessively tailored to the training dataset. Early stopping emerges as a resounding solution to address this, seamlessly intertwining with the art of model selection.

Early stopping is a form of regularization used to avoid overfitting when training learners with an iterative method, such as gradient descent. This technique halts training once a particular criterion has been met, typically when performance on a validation dataset starts to degrade.

While neural networks can fit complex datasets with high accuracy, they also tend to memorize the noise or anomalies in the training data, which harms their generalizability. By implementing early stopping, you can prevent the model from reaching this stage of excessive fitting.

Early stopping monitors a specified metric known as validation loss during the model's training. The process for using the early stopping regularization techniques is as follows:

- Train the model and periodically evaluate its performance on a validation set.
- Once the validation metric stops improving or starts worsening, wait for a predefined number of epochs, known as *"patience."*
- If no improvement occurs during the *"patience"* period, you must halt the training.

When it comes to early training in model selection, you need to know that model selection revolves around choosing the best model among a set of candidates based on their performance on unseen data. By leveraging early stopping, practitioners can identify the iteration where the model offers the optimal balance between bias and variance. This would ensure that the selected model is neither underfitting nor overfitting. Using frameworks like TensorFlow and Keras, one can easily implement early stopping in Python as follows:

```
"from keras.callbacks import EarlyStopping
# Define the early stopping criteria
early_stopping = EarlyStopping(monitor='val_loss', patience=10,
restore_best_weights=True)
# Train the model with early stopping callback
history = model.fit(train_data, train_labels, epochs=100, validation_-
data=(val_data, val_labels), callbacks=[early_stopping])"
```

In this example, training stops if the validation loss doesn't improve for ten consecutive epochs, and the model weights are reverted to those of the best epoch. However, when using this technique, you need to consider the following:

- Choice of Monitor Metric - Typically, validation loss is monitored, but other metrics like accuracy may be more suitable depending on the task.
- Setting 'Patience' - Setting an appropriate 'patience' value is crucial. A too-small value might stop training prematurely, while a too-large one could delay the detection of overfitting.
- Coupling with Other Techniques - Combining early stopping with other regularization techniques, like dropout or L1/L2 regularization, can offer compounded benefits.

Adjusting Model Complexity And Capacity

The world of neural networks offers vast potential due to its flexible architecture. However, with great flexibility comes the pivotal responsibility of sculpting the architecture. One of the cornerstones of this process is adjusting the model's complexity and capacity to ensure a balance between fitting the data well and maintaining generalization capabilities.

Model complexity is the number of parameters in your model. A more complex model has more parameters and can potentially capture more intricate patterns in the data.

On the other hand, model capacity is the model's ability to fit a variety of functions. A higher capacity model can fit more complicated functions, usually correlating with increased model complexity.

While a complex model with high capacity might seem attractive due to its ability to capture intricate patterns, it's susceptible to overfitting. Conversely, a model with too little complexity risks underfitting, failing to capture the data's underlying patterns. You must know that the model complexity can be manipulated through nodes and layers.

- **Nodes** - The building blocks of layers in a neural network. Increasing the number of nodes enhances the model's capacity.
- **Layers** - Stacked sequences of nodes. Deep networks with more layers can represent more complex functions but are more prone to overfitting.

In addition, you can also modify the model's capacity by adding nodes and layers. However, you need to increase the capacity when adding nodes, allowing the model to fit the data more closely. When you're adding layers, you need to enhance the model's depth, which can increase its symbolic power. In addition, when you're addressing the model's complexity, you need to remember the following:

- Begin with a smaller network architecture. This helps gauge a baseline performance, reducing the risk of immediate overfitting.
- Gradually increase the model's complexity by adding more nodes or layers and monitoring the performance of validation data.
- If signs of overfitting appear as the complexity increases, consider employing regularization techniques like L1, L2, or dropout to mitigate this.

On Python, you can use the TensorFlow or Keras framework to implement what we've just learned, as done below:

```
"from keras.models import Sequential
from keras.layers import Dense
# Initialize a simple model
model = Sequential()
model.add(Dense(10, input_dim=8, activation='relu')) # First hidden
layer with 10 nodes
model.add(Dense(1, activation='sigmoid')) # Output layer
# To increase complexity, add more layers or nodes:
model.add(Dense(20, activation='relu')) # Additional layer with 20 nodes"
```

However, you must consider the following when adjusting the model's complexity.

- **Computational Cost** - As complexity increases, the computational resources and time required for training may grow significantly.
- **Data Availability** - A complex model demands more data. Training a high-capacity model on limited data usually leads to overfitting.

- **Task Nature** - Some tasks inherently demand more complex models. It's pivotal to understand the problem domain and adjust the model accordingly.

Throughout this chapter, we've learned that advanced neural network architectures encompass a range of sophisticated models tailored for specific tasks and data types. These architectures have evolved from basic feedforward networks to address more complex challenges in the AI realm.

For instance, Convolutional Neural Networks (CNNs) specialize in processing grid-like data, making them ideal for image analysis. On the other hand, Recurrent Neural Networks (RNNs) are designed to handle sequential data, such as time series or natural language, by maintaining a memory of previous inputs.

There's also the rise of transformer-based models, which have set new benchmarks in language processing tasks. As data grows in complexity, these specialized architectures ensure efficient and accurate model performance across various applications.

While these models are powerful and versatile, they are not without challenges. One of the primary hurdles faced during the training phase is balancing overfitting and underfitting. Addressing these challenges requires thoughtful strategies, and one of the most effective solutions is regularization techniques. Regularization introduces constraints to the model during training, discouraging it from becoming overly complex or too simplistic.

Techniques like L1 and L2 regularization add penalties to the loss function, ensuring the weights don't grow disproportionately large. Others, like dropout, randomly deactivate specific neurons during training, promoting a more robust and generalized model.

By understanding and effectively implementing these techniques, practitioners can enhance the reliability and accuracy of their neural network models. By now, you've probably learned that data is paramount in training models. In the next chapter, we'll look at how you can prepare data for neural networks.

## 5

PREPARING DATA FOR NEURAL NETWORKS

> *"I have not failed. I've just found 10,000 ways that won't work."*

— THOMAS EDISON

Data preprocessing stands as a cornerstone in the edifice of neural network training. Neural networks, like expert artisans, require tools and materials in pristine condition to deliver their best work. In neural networks, data is both the tool and the material. This data's quality, scale, and distribution critically influence a model's accuracy, efficiency, and overall performance.

An effectively prepared dataset can easily mislead a neural network, much like a craftsman struggles with flawed tools. Feature scaling, normalization, data cleaning, and other preprocessing techniques serve to refine these tools, ensuring they are in optimal condition for training.

By standardizing the range of data features, we accelerate the training process and enhance the model's capacity to generalize from its learning. This chapter will focus on learning everything you need to prepare your data for neural networks.

UNDERSTANDING THE SIGNIFICANCE OF DATA PREPROCESSING IN NEURAL NETWORK TRAINING

As you know, well-prepared data is essential for neural networks. Feature scaling and normalization are two vital techniques you can use to prepare your data. So, with that in mind, let's explore their significance, applications, and how you can use them. By grasping the essence of these techniques, you can establish a firm foundation for training robust and high-performing neural networks.

Data Quality and Cleaning

In neural networks, data is the foundation upon which models are built. The quality of this foundation directly influences the performance and accuracy of the resulting neural network. Therefore, prioritizing data quality and ensuring proper data cleaning is not just an advisable but critical step. In addition to this, you need to understand that data quality is essential for the following reasons:

- **Influence on model performance** - no matter how sophisticated, neural networks are only as good as the data they are trained on. Poor quality data can introduce noise, leading to lower accuracy, overfitting, and poor generalization to real-world scenarios.
- **Consistency and reliability** - high-quality data ensures that the neural network's predictions remain consistent

and reliable. Inconsistent data can lead to erratic model behavior, making the model unreliable for production deployment.

- **Efficiency in training** - clean and high-quality data often reduces the time needed for training, as the model doesn't need to account for noise or irrelevant patterns.

Regarding data cleaning, one of the most common issues in datasets is missing values. Deciding whether to attribute these values, drop the rows, or use techniques like interpolation can significantly impact model performance. Let's look at these in more detail.

- **Identification** - before handling missing data, one must first identify it. This can be done using statistical tools or visualization techniques.
- **Imputation** - once identified, missing data can be addressed through various imputation methods. Mean or median imputation, K-nearest neighbors, and even using smaller neural networks or regression models are standard techniques.
- **Deletion** - in cases where the proportion of missing data is negligible, deleting the affected rows or columns might be more efficient. However, one should exercise caution to avoid introducing bias.

In addition, you also need to know that methods like using regression, model-based imputation, or even unsupervised neural networks can be leveraged to predict and fill in missing values, ensuring a higher degree of accuracy.

Another thing you need to consider when clearing the data is Identifying and Removing Outliers. They can distort the model's understanding of data patterns. Visual methods like box plots or mathematical methods such as the IQR (Interquartile Range) technique can help identify outliers.

- **Visualization** - techniques such as scatter plots, box plots, or histograms can be instrumental in spotting outliers.
- **Z-Score or IQR** - statistical methods like the Z-Score or the Interquartile Range (IQR) method can help systematically identify and remove outliers.

Once identified, they can be transformed or removed to prevent skewing the training process. Data collected from multiple sources or over different periods might be inconsistent, and you need to ensure its conformity. Standardizing terminologies, units, and scales is essential. For example, ensuring that all temperature readings are in Celsius rather than a mix of Fahrenheit and Celsius.

It would be best if you also focused on removing duplicate rows. They add no value but increase the size of your dataset, wasting computational resources. Duplicate data can lead to overfitting, as the model might see the same data multiple times during training. Tools and libraries like pandas in Python offer straightforward methods to identify and remove duplicate records. To optimize and remove duplicate rows, you can use the following function:

"drop_duplicates()"

Let's walk through a simple example in Python using the Pandas library, commonly used for data manipulation and cleaning tasks.

This example will encompass some of the data-cleaning steps mentioned earlier:

```
"# Import necessary libraries
import pandas as pd
import numpy as np
# Sample data: Imagine a dataset of ages and salaries, but it contains
missing values, duplicates, and outliers.
data = {
'Age': [25, 30, np.nan, 45, 50, 25],
'Salary': [50000, 55000, 60000, 2000000, 65000, 50000] # Notice the
outlier in Salary
}
df = pd.DataFrame(data)
# 1. Handling Missing Values:
# Impute missing age with the median of the Age column
df['Age'].fillna(df['Age'].median(), inplace=True)
# 2. Removing Outliers:
# Here, we'll handle outliers for the Salary column using the IQR method
Q1 = df['Salary'].quantile(0.25)
Q3 = df['Salary'].quantile(0.75)
IQR = Q3 - Q1
lower_bound = Q1 - 1.5 * IQR
upper_bound = Q3 + 1.5 * IQR
# Filter out the outliers
df = df[(df['Salary'] >= lower_bound) & (df['Salary'] <= upper_bound)]
# 3. Standardizing Data:
# Using Z-score standardization for the Age column as an example
mean_age = df['Age'].mean()
std_age = df['Age'].std()
df['Age'] = (df['Age'] - mean_age) / std_age
# 4. Handling Duplicate Rows:
```

df.drop_duplicates(inplace=True)
print(df)"

As you can see in the example above, we first handle missing values in the *"Age"* column by filling them with the median age. Once that's done, we address outliers in the "Salary" column using the IQR method. In the example above, the *"Age"* column is standardized using Z-score standardization, and when we enter the function, the duplicate rows are removed.

Ensuring quality and cleanliness is paramount before introducing data to a neural network. The data acts as the primary source of information for the model, and any misrepresentation can lead to inefficient and erroneous outputs. Devoting time to data preprocessing isn't just a preliminary step; it's a strategic move that dictates the success of your neural network's performance.

Feature Scaling And Normalization

Regarding data preparation, feature scaling, and normalization are two essential data preprocessing steps in neural network training. These steps ensure that the data used to train a model is consistent and can be processed effectively. Using these techniques ensures a more efficient training process and a model with a better generalization capability.

Before we proceed, let's look at these techniques individually to understand them comprehensively. Feature scaling is a technique used to standardize the range of independent variables or features of data. In data processing, it is also known as data normalization and is essential for a variety of reasons, including:

1. **Consistent influence** - if you're not using feature scaling, a variable with a more extensive range might overshadow another variable with a smaller range, leading to misleading results.
2. **Improved gradient descent convergence** - when you use feature scaling, you ensure that the optimization algorithm converges more quickly. This leads to faster training times for neural networks.
3. **Enhanced model performance** - you need to know that a neural network can struggle to capture relationships when input features have different scales. By standardizing the scales, you can ensure that the model quickly identifies patterns in the data.

On the other hand, normalization is often called Min-Max scaling. In this technique, transform features to fall within a specified range. A typical example of that range would be [0, 1]. It ensures that each feature has the same scale, making it easier for optimization algorithms to converge. Normalization is essential for various reasons, including:

- **Uniformity in features** ensures that each feature contributes equally to the distance calculations and model's decisions.
- **Beneficial for algorithms requiring distance computations** - you must know that algorithms like k-NN, k-Means, and logistic regression can converge faster when data is normalized.
- **Ensures data consistency** - normalization brings consistency to the dataset when dealing with features measured in different units.

Both feature scaling and normalization are vital preprocessing steps to ensure a neural network performs optimally. While the techniques may seem similar, their applications and impacts can vary based on the nature of the data and the problem at hand. Understanding when to use each method and its implications on the modeling process is crucial.

You need to know that feature scaling standardization using the Z-score can be used in various cases that include:

- Principal Component Analysis (PCA).
- Support Vector Machines (SVM).
- K-Means Clustering.

PCA is a dimensionality reduction technique that transforms high-dimensional data into a new coordinate system. Before applying PCA, it's essential to standardize the data since PCA is sensitive to variances. Standardized data ensures that each feature contributes equally to the principal components.

SVM is a supervised machine learning algorithm used for classification or regression. It tries to find the best hyperplane that separates the classes. You must know that SVM's optimization process requires standardized data to ensure each feature influences the model uniformly. Large-scale features could dominate the model without standardization, leading to suboptimal hyperplanes.

K-Means is an unsupervised clustering algorithm that classifies data points into "K" number of clusters. K-Means uses distance metrics to compute similarity. Standardized data ensures that each feature's scale doesn't distort these distance calculations. Thus, clusters formed are more accurate and meaningful. Normalization, on the other hand, can be used for:

- Neural Networks.
- Gradient Descent-Based Algorithms.
- Image Processing.

Neural networks are computational models inspired by the human brain's structure. They are used for classification, regression, and even complex tasks like image and voice recognition. You must know that neural network algorithms often require input data within a specific range, typically [0, 1], especially when using activation functions like the sigmoid or tanh. Normalizing the input data improves the network's convergence and performance.

Gradient descent is an optimization algorithm used to minimize a function iteratively. Algorithms like linear regression that employ gradient descent benefit from normalization as it helps the algorithm converge faster. The contour plot is more spherical when features are on the same scale, leading to fewer oscillations and quicker convergence.

In image processing, tasks like image recognition, classification, and enhancement are done daily. Regarding processing, you must know that the image pixel values typically range from 0 to 255. Normalizing these pixel values to the range [0, 1] is common, especially when feeding images to machine learning models. This ensures consistent input data and often leads to better model performance.

Both feature scaling and normalization play vital roles in specific scenarios. The choice between them, or even the choice to use both, depends on the nature of the dataset and the requirements of the specific algorithm in use. If you want to implement feature scaling in Python, you'll need to use the *"sklearn.preprocessing"* library, and here's how the implementation would play out:

```
"from sklearn.preprocessing import StandardScaler
import pandas as pd
# Sample data
data = {
'Age': [25, 30, 35, 40, 45],
'Salary': [50000, 55000, 60000, 65000, 70000]
}
df = pd.DataFrame(data)
# Z-Score Normalization (Feature Scaling)
scaler = StandardScaler()
df_scaled = scaler.fit_transform(df)"
```

You need to know that you can also use the *"sklearn.preprocessing"* In Python for normalization, and the source code for that would be similar to the one provided below.

```
"from sklearn.preprocessing import MinMaxScaler
import pandas as pd
# Sample data
data = {
'Age': [25, 30, 35, 40, 45],
'Salary': [50000, 55000, 60000, 65000, 70000]
}
df = pd.DataFrame(data)
# Min-Max Scaling (Normalization)
minmax_scaler = MinMaxScaler()
df_normalized = minmax_scaler.fit_transform(df)"
```

By comprehensively understanding and implementing feature scaling and normalization, you ensure that the foundational elements of your neural network training are vital, paving the way for a robust and high-performing model. Now, let's look at some beginner-friendly techniques to help you prepare your data for a neural network.

SIMPLIFYING DATA CLEANING, NORMALIZATION, AND FEATURE SCALING TECHNIQUES FOR BEGINNERS

In real-world scenarios, datasets often present irregularities such as missing values and outliers. Addressing these irregularities is crucial because, if overlooked, they can compromise the integrity of a neural network's training process and hinder its performance. However, various techniques can be used to address such concerns.

Handling Missing Data And Outliers

Before we proceed with how to handle missing data, you need to comprehend why data would be missing in the first place. Data can go missing for various reasons. Understanding these reasons can aid in selecting the appropriate method for handling such omissions. Sometimes, data goes missing purely by chance, without any systematic cause. For instance, a malfunctioning sensor might sporadically fail to record data.

Sometimes, the data might be consistently missing due to an underlying pattern or reason. For example, older entries in a dataset might lack specific attributes introduced only in recent records. Another reason some data might be missing is the data collection process. During the data collection, errors such as instrument malfunctions, misreporting, or oversight by data collectors can lead to missing data.

In addition, data might be intentionally withheld or not reported. For instance, respondents in a survey might choose not to answer specific sensitive questions. Another common cause of missing data is merging data sets. When merging datasets from different sources or formats, specific attributes might not align perfectly, leading to missing values.

Before addressing concerns about missing data, you also need to know what outliners are. These are data points that differ significantly from other observations in the dataset. Their presence can be the result of various factors that include:

1. **Genuine variations** - outliers sometimes reflect actual variations in the data. For example, in a dataset of employees' salaries, the CEO's salary might be an outlier compared to others.
2. **Data entry errors** - human error during data input, such as typographical mistakes, can lead to outliers. For instance, adding an extra zero to a figure can drastically change its value.
3. **Measurement errors** - outliers can result from faults in the data collection process. A faulty sensor, for example, might record abnormally high temperatures.
4. **Intentional outliers** - sometimes, data points are deliberately manipulated to appear as outliers, especially in fraud detection cases.
5. **Natural discrepancies** - not all data conform to our expectations or standard patterns in real-world scenarios. Rare events or circumstances can lead to natural outliers in datasets.

Recognizing the root causes of missing data and outliers is pivotal. It helps in effectively addressing these issues and gaining a more profound understanding of the underlying data structure and its intricacies. Proper handling and interpretation of these irregularities ensure that the subsequent data analysis or modeling stages are based on a reliable foundation.

Strategies For Handling Missing Data

When it comes to handling missing data, there are different techniques you can use. However, before we get into what these techniques are to you, what they are suitable for may vary. In addition, some of the techniques may come with certain drawbacks. Some of the most straightforward strategies for handling missing data include:

1. Deletion - the most straightforward approach is to remove any row containing missing data. However, this can lead to the loss of valuable information, especially if a substantial amount of data is missing. You can implement this in Python using:

"import pandas as pd
df.dropna(inplace=True)"

2. Imputation - as mentioned, imputation can be further broken down into three categories that include:

 a. Mean or median imputation - replaces missing values for normally distributed or skewed data. In Python, this would look like:

"# For mean
df.fillna(df.mean(), inplace=True)
 # For median
df.fillna(df.median(), inplace=True)"

b. Mode imputation - useful for categorical data. Replace missing values with the most frequent value. This can be implemented in Python using the following:

"df.fillna(df.mode().iloc[0], inplace=True)"

c. Predictive imputation - allows you to use ML models to predict and replace missing values based on other features. In Python, you can fill in the missing values as done. For the example below, *"-9999"* is the missing value.

"df.fillna(-9999, inplace=True)"

Strategies For Handling Outliers

As mentioned, outliers are data points that diverge from other observations. You must know that while some outliers can be genuine, indicating a new or rare event, others can be errors. Some of the most straightforward strategies for handling outliers include:

1. Box plot method - a box plot, also known as a whisker plot, displays the summary of a set of data values. Points outside the whiskers are considered outliers. To use this method, what you have to do in Python is mentioned below:

"import matplotlib.pyplot as plt
plt.boxplot(df['Feature_Name'])
plt.show()"

2. Z-Score method - the Z-score represents how many standard deviations a data point is from the mean. A high absolute value of

Z-score indicates it's an outlier. It can be implemented in Python using the following:

"from scipy import stats
z_scores = stats.zscore(df)
abs_z_scores = abs(z_scores)
filtered_entries = (abs_z_scores < 3).all(axis= 1)
df = df[filtered_entries]"

3. IQR method - the Interquartile Range (IQR) is the difference between the 75th percentile (Q3) and the 25th percentile (Q1). Any data point outside the range [Q1 - 1.5*IQR, Q3 + 1.5*IQR] is an outlier. It can be executed in Python as follows:

"Q1 = df.quantile(0.25)
Q3 = df.quantile(0.75)
IQR = Q3 - Q1
*df = df[~((df < (Q1 - 1.5 * IQR)) | (df > (Q3 + 1.5 * IQR))).any(axis= 1)]"*

By effectively understanding and addressing missing data and outliers, you can lay a robust foundation for your data prepro-cessing efforts. This will allow you to ensure that your neural network models are built on reliable and high-quality data. Proper handling of these issues leads to better model performance and provides clearer insights into the underlying patterns and rela-tionships within the data.

Scaling Numerical Features And One-Hot Encoding Categorical Variables

Scaling refers to transforming numerical features to standardize their range or distribution. Adjusting the scale ensures that no particular feature disproportionately influences the algorithm due to its range or unit. Scaling numerical features is essential for several reasons: uniformity, performance, and interpretability.

Algorithms, particularly those that rely on distances like k-means clustering or gradient descent, work more efficiently when all features have a consistent scale. This allows them to have faster convergence. It's also important to know that a feature with a larger scale might unduly dominate an algorithm's behavior. Scaling ensures each feature has an equitable influence based on its merit.

Another thing that you need to remember is that with features on a similar scale, coefficients or feature importance values in machine learning models become more interpretable. There are several methods for scaling, but two common techniques are Min-Max scaling and Standard scaling (or Z-score normalization). Let's look at each one in more detail.

1. Min-max scaling - this transforms features by scaling them to a given range, often [0, 1]. This method can be used in Python as follows:

"from sklearn.preprocessing import MinMaxScaler
scaler = MinMaxScaler()
df[['numerical_feature']] =
scaler.fit_transform(df[['numerical_feature']])"

2. Standard scaling - this method centers the feature around zero and scales based on the standard deviation. You can use this method in Python as follows:

```
"from sklearn.preprocessing import StandardScaler
scaler = StandardScaler()
df[['numerical_feature']] =
scaler.fit_transform(df[['numerical_feature']])"
```

On the other hand, one-hot encoding is a process of converting categorical data variables into a format that machine learning algorithms can use. It involves representing each definite value as a unique binary vector. This is essential for various reasons that include:

- Compatibility with algorithms.
- Model performance.
- Avoidance of ordinal assumptions.

Before we dive into how this method can be used, you need to know that most ML algorithms require numerical input. One-hot encoding transforms categorical data into a machine-readable format without introducing ordinal relationships that aren't present in the original data.

In addition, when you represent categories as orthogonal vectors, algorithms can more easily discern the distinct category effects. Lastly, you also need to know that label encoding, another common technique, might introduce ordinal relationships where none exist. However, this can be avoided using one-hot encoding.

To one-hot encode categorical variables in Python, you can use the *"pandas"* library as mentioned below:

```
"import pandas as pd
df_encoded = pd.get_dummies(df, columns=['categorical_feature'],
drop_first=True)"
```

In the above example, *"drop_first=True"* is used to avoid the "dummy variable trap", which can cause multicollinearity in regression models. The method is also essential since most algorithms need numerical input. One-hot encoding ensures categorical data is represented in a machine-friendly format.

In addition, unlike other encoding methods that might introduce ordinal interpretations, one-hot encoding preserves the categorical nature without implying a rank. Data scientists can lay a robust foundation for building accurate and reliable ML models by ensuring numerical features are appropriately scaled and categorical variables are efficiently encoded.

Although these techniques are highly beneficial, they do come with a set of unique challenges. Some common challenges that you can experience with scaling features include:

- **Loss of original data distribution** - while scaling alters the range or distribution of the data, it's pivotal to remember that the original distribution is modified. Sometimes, having features in their original scale can provide valuable domain-specific insights that might get obscured post-scaling.
- **Vulnerability to outliers** - techniques like Min-Max scaling are susceptible to outliers. A single extreme outlier can skew the scaled range for all other values, causing a loss of variability in the data.
- **Assumption of linearity** - by scaling data, there's an implicit assumption that relationships between variables are linear. Scaling might not always be beneficial for algorithms where this isn't a prerequisite.
- **Maintaining consistency during deployment** - the scaling parameters (minimum and maximum values for

Min-Max scaling or mean and standard deviation for Standard scaling) determined during the training phase must be consistently applied during deployment on new data. Ensuring this consistency can pose challenges, especially in dynamic environments.

You can also run into some roadblocks in one-hot encoding. Some of the most common challenges include:

- **Dimensionality increase** - one-hot encoding can drastically increase the number of features, especially with high-cardinality categorical variables. This surge in dimensionality might lead to dimensionality, making models more complex and potentially leading to overfitting.
- **Memory overhead** - with a significant increase in columns, memory consumption can rise, making it challenging to handle large datasets. This might require more sophisticated data storage and processing solutions.
- **Lack of ordinal representation** - for genuinely ordinal data, one-hot encoding might not capture the inherent order. Alternatives like ordinal encoding would be more appropriate, though they have challenges.
- **Handling new categories in deployment** - if a new category not present during training is introduced in the live data, it can pose challenges. Ensuring that the model can handle or ignore new categories becomes essential.

Effective data preprocessing is imperative for the successful application of neural networks. Throughout this chapter, we've emphasized the importance of various preprocessing steps, from managing missing data and outliers to appropriately scaling and encoding features. These processes ensure optimal model performance by providing a consistent and representative data structure.

It would be best to remember that as you go deeper into the complexities of neural network training, using a meticulous approach for data preparation will become paramount. Implementing the practices we've discussed enhances model accuracy and ensures robustness against varied data scenarios. Now that you know how to prepare data for neural networks, we'll focus on determining how you can make the networks more Interpretable in the next chapter.

## 6

MAKING NEURAL NETWORKS INTERPRETABLE

"Technology like art is a soaring exercise of the human imagination."

— DANIEL BELL

Interpreting neural networks can be challenging; however, it is essential for practical, real-world applications. The critical task of peeling back the layers of neural network models to understand how a model makes decisions requires different techniques. In this chapter, we'll focus on understanding the nature of neural networks.

Then, we'll dive into different techniques for understanding the model's behavior. We'll focus on visualization techniques like filters and maps and will also cover plotting learning curves, feature attributes, and saliency maps. So, with that in mind, let's get started.

ADDRESSING THE CHALLENGES OF INTERPRETING NEURAL NETWORK DECISIONS FOR BEGINNERS

Before diving into how you can make neural networks interpretable, you need to understand their nature and what makes interpreting them a challenge. It's crucial for those entering the field to grasp why these models, despite their power, often operate as black boxes.

This section will break down the critical need for interpretability across various domains and its impact on trust, legal compliance, and ethical AI use. Doing so will help you understand why making neural network decisions transparent and accountable is a fundamental step.

The Black-box Nature Of Neural Networks

The term *"black-box"* in the context of neural networks refers to systems where the internal workings are either unknown to the observer or are known but not understandable. A neural network becomes a black-box when we can observe its inputs and outputs but cannot explain the process of getting from one to the other with clarity and transparency. You need to know that this can happen for a variety of different reasons, including:

- Complexity architecture
- Parameter volume
- Opaque data representations
- Interdependent features
- Data learning
- Non-intuitive optimization

The pressing need to address the black-box nature of neural networks arises from a fundamental requirement for trust, reliability, and ethical use of AI. For instance, when a neural network is used to approve loan applications, its decision affects the lives of real people.

If the system denies an application, the applicant deserves an explanation, which the black-box model may fail to provide. This lack of transparency can conceal flawed reasoning, systematic biases, or other significant deficiencies in the model. Let's look at these factors in more detail.

Lack of Transparency And Understanding

You need to know that transparency in neural networks means the ability to see and comprehend the inner workings of the models. A significant barrier to transparency is the intricate complexity within deep learning architectures. These networks often consist of thousands to millions of parameters, which interact in highly non-linear ways.

As a result, the relationship between input data and prediction is not transparent, making it challenging for even the model's developers to understand the specific reasons behind a given output.

Non-intuitive Features And Patterns

Neural networks, especially deep learning models, can capture non-intuitive features and patterns that are not evident to humans. They can identify and leverage correlations in the data that may be beyond human recognition or seem irrelevant to the task.

For example, a network trained to recognize objects in pictures might fixate on a background pattern that coincidentally appears in the images of a particular class rather than the object itself.

Limited Explainability

Explainability is a measure of how well the internal mechanics of a system can be described in human-understandable terms. Neural networks' limited explainability arises from their complex data transformations and layering, which defy simple explanations.

For instance, while we can articulate the function of individual neurons in a single-layer perceptron, the same cannot be said for deep neural networks where subsequent layers abstract the data representations progressively.

Potential Bias And Discrimination

You need to know that neural networks, by design, are susceptible to encoding biases in their training data. Understanding that these biases can lead to discriminatory practices when deployed in the real world is essential.

For instance, a facial recognition system trained predominantly on images of one ethnic group will likely perform poorly on others. Such biases can perpetuate and even exacerbate existing societal prejudices if left unchecked.

Trust And Accountability Concerns

For neural networks to be fully integrated into critical decision-making processes, they must be deemed trustworthy. Trust comes from understanding and the ability to predict a system's behavior in various situations, which the black-box nature undermines.

Additionally, the question of accountability arises when a decision made by an AI system has adverse consequences. Without interpretability, it becomes challenging to pinpoint responsibility for these outcomes.

Efforts To Improve Interpretability And Explainability

The field of explainable AI (XAI) is devoted to bridging the gap between the performance of neural networks and the human understanding of their decision-making processes. Efforts include developing visualization techniques that can illustrate the activation of neurons in response to inputs, creating models that can report the importance of different features in their predictions, and designing networks with inherent interpretability.

By enhancing the interpretability and explainability of neural networks, researchers and practitioners aim to build systems that are powerful in their predictive capabilities and accountable, trustworthy, and free of biases.

The Need For Interpretability In Real-World Applications

In deploying neural networks within real-world applications, interpretability is not just something nice to have. It's a critical component that can determine the success or failure of the system. Some of the main reasons why this is important include:

- **Regulatory Compliance**

Regulatory bodies may require automated decision-making systems to be transparent and accountable. These regulations necessitate that organizations be capable of elucidating the inner workings of their AI models. This means that It is critical for entities employing neural networks to ensure they can clearly articulate the rationale behind model decisions.

- **Trust and Acceptance**

For users and decision-makers to trust AI systems, they need transparency. Understanding how a model makes decisions builds confidence in its reliability and ensures its broader acceptance. This is especially crucial in sectors like healthcare and finance, where decisions directly affect human lives.

- **Error Detection and Debugging**

Another reason this is important is because interpretability allows developers to examine the decision-making process of a neural network, making it easier to identify and correct errors. Diagnosing the root cause of failures or unexpected behavior can be challenging without it.

- **Model Improvement and Refinement**

Interpretable models enable a feedback loop for continuous improvement. By understanding how different features influence the outcome, developers can refine their models for better performance. This may include taking measures such as re-engineering features, reselecting or reweighting inputs, or even restructuring the model architecture.

- **Fairness and Bias Mitigation**

An interpretable model lays bare the criteria it uses to make decisions, allowing for the detection and mitigation of biases. This is vital for ensuring that AI systems do not perpetuate or amplify societal inequalities and treat all individuals and groups fairly.

- **Risk Assessment and Decision Support**

In high-stakes environments, such as autonomous driving or medical diagnosis, understanding the reasons behind a decision is as important as the decision itself. Interpretability provides insight into the potential risks and helps create robust decision support systems that can collaborate effectively with human experts.

- **Ethical Considerations**

As AI systems become more autonomous, the ethical implications of their decisions grow. Interpretability ensures these systems operate within ethical boundaries and align with human values and moral principles.

- **Human-AI Collaboration**

Interpretability enhances the collaboration between humans and AI, enabling humans to understand, predict, and effectively manage AI behavior. This collaboration can improve outcomes in clinical settings where AI supports medical professionals in diagnosis and treatment planning.

- **Real-World Impact**

The ultimate test of any technology is its impact on the real world. For AI systems to have a positive effect, they must be interpretable to be monitored, evaluated, and integrated into society in a way that aligns with public interest and welfare.

INTRODUCING VISUALIZATION TECHNIQUES FOR UNDERSTANDING MODEL BEHAVIOR

Visualization techniques are essential for understanding model behavior. They transform abstract data into intuitive graphical representations, enabling us to scrutinize a model's internal mechanisms. By visualizing activation maps, filters, and feature maps, especially within Convolutional Neural Networks (CNNs), we gain insight into the learned features and their influence on the network's decision-making process.

Moreover, plotting learning curves and loss landscapes reveals crucial dynamics of the training phase, helping to pinpoint issues like overfitting or undertraining. These visualization tools don't just add transparency; they are vital for refining and validating neural network architectures.

Visualizing Activation Maps And Filters In CNNs

CNNs are particularly adept at handling visual tasks, but understanding how they discern and process optical inputs can be challenging. Visualization techniques such as activation maps and filters become instrumental in demystifying the internal mechanisms of CNNs. Before we talk about doing this in Python, there are a few things you need to know.

Filters

Filters, or kernels, are the core components of a CNN that perform the convolution operation across an image. Each filter detects a specific feature from the input. Visualizing filters helps in comprehending what kind of features the network is sensitive to. In the early layers, filters might be simple edge or color detectors.

As we go deeper, filters combine these primary features to detect more complex patterns. One common approach to visualize these filters is optimizing an input image so that a particular filter is activated. This effectively shows what the filter is looking for. You can also directly examine the weights of the filters, which can be plotted as images.

These visualizations often reveal intricate patterns when we reach the deeper layers of the network. Using filters allows you to test and demonstrate the complexity of features the network has learned to recognize.

We can access the filter weights and plot them to visualize a filter. The filters in the initial layers of a CNN will look like small patches, showing simple patterns. As you move deeper into the network, these filters become more abstract and more complex to interpret because they represent higher-level features. Provided below is an example of how you can visualize filters in Python.

```
"from keras.models import Model
from matplotlib import pyplot as plt
# Assume 'model' is a Keras model you've already created.
# Retrieve the weights from the first convolutional layer
filters, biases = model.layers[1].get_weights()
# Normalize filter values between 0 and 1 for visualization
f_min, f_max = filters.min(), filters.max()
filters = (filters - f_min) / (f_max - f_min)
# Plot first few filters
n_filters = 6
for i in range(n_filters):
# Get the filter
f = filters[:, :, :, i]
# Plot each channel separately
```

```
for j in range(3): # Assuming RGB channels
plt.subplot(n_filters, 3, i * 3 + j + 1)
plt.imshow(f[:, :, j], cmap='gray') # Can adjust cmap for better visual-
ization
plt.axis('off')
plt.show()"
```

Activation Maps

Activation maps provide insight into which features of an input image activate specific neurons in the network. By visualizing these activations, we can interpret which aspects of the data are highlighted at each layer of the CNN. When an image is passed through a CNN, the activation maps evolve, presenting more abstract representations.

Activation maps might initially highlight basic features like edges and colors. However, as we progress through layers, they represent more complex structures, such as textures and patterns specific to the task at hand, such as the shape of an object. The technique to visualize these activations is straightforward: pass an image through the network.

After that, capture the output of each convolutional layer and plot the resulting activation maps. This process reveals the successive transformation of the input and provides an understanding of how the network interprets the image.

We need to run an input image through the network to visualize the activation maps and collect the outputs from the convolutional layers. Then, these outputs can be plotted to show what the model is seeing. An example of how activation maps can be visualized in Python is provided below:

```
"from keras.models import Model
# Define a new model that will take an image and output the activation
maps
activation_model = Model(inputs=model.input, outputs=[layer.output for
layer in model.layers if 'conv' in layer.name])
# Assume 'img' is the preprocessed image ready to be fed to the model
activations = activation_model.predict(img)
# Plot the activations of the first conv layer for the first image
first_layer_activation = activations[0]
plt.figure(figsize=(10, 10))
for i in range(first_layer_activation.shape[-1]): # Iterate over all the
filters
plt.subplot(6, 6, i+1) # Arrange images in 6x6 grid
plt.imshow(first_layer_activation[0, :, :, i], cmap='viridis') # Use a
colormap that highlights features
plt.axis('off')
plt.tight_layout()
plt.show()"
```

PLOTTING LEARNING CURVES AND LOSS LANDSCAPES

Understanding the behavior of neural networks during training is crucial for model development and refinement. Therefore, you need to know that visualization techniques such as plotting learning curves and loss landscapes are instrumental in this process. These not only provide clear, interpretable metrics of model performance over time.

Learning Curves

Learning curves plot training and validation loss or accuracy over iterations or epochs, providing immediate visual feedback on how well a model learns from the data. Some of the critical insights you can gain from plotting learning curves include:

- **Training progress** - a decreasing trend in the training loss indicates that the model is learning as expected.
- **Generalization gap** - the training and validation metrics gap suggests how well the model generalizes to unseen data. A small gap indicates good generalization, while a large gap may point to overfitting.
- **Convergence** - the point at which the plateau of the curve suggests that the model has learned as much as it can and additional training would not lead to significant improvements.

When interpreting learning curves, you must look for decreasing training loss and gaps between curves and plateaus. It's important to know that the training loss should decrease steadily. Any sharp drops or increases may indicate learning rate or data quality issues.

In addition, a consistent gap between training and validation curves may suggest overfitting, where the model performs well on the training data but poorly on validation data. Furthermore, plateaus can indicate a learning slowdown, while oscillations may suggest that the learning rate is too high.

You can adjust learning curves by early stopping, optimizing learning rates, and using regularization techniques. However, you need to know how these techniques should be used. Early Stopping should be used if validation metrics worsen while training metrics improve; it might be time to stop training to prevent overfitting.

You can optimize learning rates to help avoid oscillations or plateaus in the learning curves.

In addition, you can implement methods like dropout or L2 regularization to reduce overfitting.

The accuracy at each epoch during training and validation can be calculated as the number of correct predictions divided by the total number of predictions.

Accuracy=Total Number of Predictions/Number of Correct Predictions

The loss for a set of predictions can be calculated using a loss function, such as the mean squared error for regression problems:

$$MSE = n \, 1\sum i = 1n(yi - y\hat{}i)2$$

Loss Landscapes

Loss landscapes visualize the behavior of the loss function concerning the model parameters. This visualization helps understand the optimization challenges by showing where the loss function has the minima, maxima, and saddle points. A loss landscape can reveal:

- **Convexity** - ideally, loss landscapes should be convex; however, they are riddled with complexities that can trap optimization algorithms.
- **Local minima** - these are points where the loss is lower than in the immediate vicinity but not necessarily the lowest point of the function.

- **Saddle points** - points where the loss is not a local minimum or maximum but where the gradient is zero can slow down training.

Two main benefits of visualizing loss landscapes include optimization monitoring and detection of overfitting. Optimized monitoring allows you to understand the topology of the loss landscape and can help diagnose why a model might be struggling to learn. In addition, detecting overfitting in the loss landscape can indicate a model that memorizes the training data rather than learning general features.

You need to know that visualizing the loss landscape helps interpret the neural network and make necessary adjustments. By visualizing the loss landscape, we can get a sense of the difficulty of the optimization process. Observing sharp valleys or ridges can guide our decisions on adjusting the optimization algorithm or learning rate schedules.

Visualization tools such as Matplotlib for static plots or TensorBoard for interactive visualizations facilitate the plotting of learning curves and loss landscapes. They are part of an iterative process where the model is evaluated, issues are identified, and improvements are guided based on observed patterns.

Incorporating these visualization techniques into the model development workflow enables a more systematic and informed neural network training and evaluation approach. By observing and interpreting these visual clues, we can make informed decisions to fine-tune our models for optimal performance.

The loss surface of a neural network can be represented as a high-dimensional surface, where the loss is a function of the network's parameters θ:

$$L(\theta)$$

However, visualizing this high-dimensional space requires dimensional reduction techniques or plotting slices of the space by varying only a couple of parameters while keeping others constant. A common technique for visualizing loss landscapes is to plot the loss contour, which is a 2D representation, where two parameters θ_1 and θ_2, are varied, and the loss is plotted as the z-axis:

$$L(\theta_1, \theta_2)$$

These formulas are the backbone of the computations that go into plotting the performance metrics over time (learning curves) and visualizing how changes in the model parameters affect the loss function (loss landscapes). Advanced visualization tools then help interpret these plots, allowing you to make informed decisions about the training and architecture of your model.

FEATURE IMPORTANCE AND SALIENCY MAPS

Feature importance and saliency maps are critical techniques for interpreting neural network decisions. They provide clear indicators of influential data points and guide model refinement for more accurate predictions. Let's look at them in more detail.

Feature Attribution Methods: Gradient-based Approaches

Feature attribution methods are critical for the interpretability of neural networks, especially to explain the predictions of deep learning models. Gradient-based approaches stand out due to their effectiveness and mathematical grounding.

These methods attribute the prediction of a neural network to its input features based on the gradients of the output concerning the input. Gradient-based techniques, such as Gradient-weighted Class Activation Mapping (Grad-CAM), use the gradients flowing into the final convolutional layer to understand which features activate specific neurons and impact the output prediction.

The intuition is that if a slight change in a feature significantly affects the output, this feature is essential for the prediction. The process involves computing the gradient of the class score concerning the feature maps of a convolutional layer. This indicates the spatial locations that most strongly activate for a given class.

These gradients are then pooled to derive the neuron importance weights. By overlaying the heatmap generated from these weights onto the original input image, practitioners can visualize the areas of the image that contributed most to the model's decision. This is invaluable for tasks such as object detection and image classification. This helps understand the model's focus and can reveal potential biases or errors in the training data.

Extensions of this approach, like Integrated Gradients and Layerwise Relevance Propagation (LRP), offer even more insights. For instance, integrated gradients measure each pixel's contribution by integrating the gradient along the path from a baseline (an input that represents no information) to the actual input. This overcomes some of the limitations of simple gradient methods.

These gradient-based methods are more than diagnostic tools; they bridge human understanding and the often opaque reasoning of complex models. Implementing these techniques allows for a more responsible deployment of AI, ensuring that model decisions can be explained and justified meaningfully. Below is an example of how you can do this in Python:

```
"import tensorflow as tf
# Assuming 'model' is a trained Keras model and 'input_img' is the input
image tensor
# Function to compute gradients
def compute_gradients(model, input_img, target_class_idx):
with tf.GradientTape() as tape:
tape.watch(input_img)
predictions = model(input_img)
loss = predictions[0][target_class_idx]
# Compute gradients concerning input image
gradients = tape.gradient(loss, input_img)
return gradients
# Function for Integrated Gradients
def integrated_gradients(model, input_img, baseline_img, target_-
class_idx, steps=50):
# Scale input images and compute gradients
scaled_inputs = [baseline_img + (float(i) / steps) * (input_img - base-
line_img) for i in range(0, steps + 1)]
grads_list = [compute_gradients(model, scaled_input, target_class_idx)
for scaled_input in scaled_inputs]
# Average the gradients
average_grads = tf.reduce_mean(tf.stack(grads_list), axis=0)
# Compute the integrated gradients
integrated_grads = (input_img - baseline_img) * average_grads
return integrated_grads
# Prepare a baseline image (could be a zero image or a dataset mean)
baseline_img = tf.zeros_like(input_img)
# Choose the target class index (assuming you know the target class)
target_class_idx = 123 # Example class index
```

```
# Compute Integrated Gradients
attributions = integrated_gradients(model, input_img, baseline_img,
target_class_idx)
# Visualize the attributions
# Depending on the data, you may plot them as a heatmap, overlay them,
etc."
```

Visualizing Saliency Maps For Input Attribution

Visualizing saliency maps is a powerful technique to attribute the input features contributing to the neural network's predictions. Saliency maps highlight the most influential parts of the input data, such as pixels in an image, by showing where the model is looking to make a decision.

Creating a saliency map typically begins by computing the output gradient concerning the input image. This gradient reflects how changes in input pixel intensity affect the change in output prediction. When these gradients are calculated for a specific output class, they can be used to generate a map that visually represents the importance of each pixel in contributing to the final decision.

You need to know that a saliency map can be generated by taking the absolute value of the gradient for each input pixel, which results in a heatmap. Higher intensities in this heat map correspond to pixels that substantially influence the model's output more. It is essential to apply proper normalization to ensure that the heatmap is visually interpretable.

These maps reveal which features are most important and highlight whether the network is considering irrelevant features due to overfitting or biases in the training data. These insights are precious for debugging and improving the model's performance.

They provide a direct visualization of which parts of the input data are weighted heavily by the network. Moreover, saliency maps serve as an aid in enhancing trust and transparency in AI systems. By clarifying which aspects of the data are pivotal in a model's decision-making process, you can better understand and trust the model's predictions.

However, it is also important to note that standard saliency map techniques have limitations. They can sometimes produce noisy visualizations or may not fully capture the complexity of the model's decision-making process. Researchers have developed variants and improvements to address these issues, such as SmoothGrad.

This reduces noise by averaging the gradients of multiple inputs with added noise, thereby creating smoother and more inter-pretable maps. To generate a saliency map in Python, you need to:

1. Load the pre-trained neural network model that you want to interpret.
2. Format the input data to be compatible with the model. This usually involves resizing, normalizing, and possibly expanding the dimensions to include a batch size.
3. Calculate the gradients of the output concerning the input image using backpropagation.
4. Process the computed gradients to generate a saliency map.
5. Visualize the saliency map by overlaying it on the input image to show which parts strongly influence the model's predictions.

Below is an example of how to generate saliency maps in Python using Keras and TensorFlow.

```
"from tensorflow.keras.applications import VGG16
from tensorflow.keras.preprocessing import image
from tensorflow.keras.applications.vgg16 import preprocess_input
import numpy as np
import tensorflow as tf
import matplotlib.pyplot as plt
# Load a pre-trained VGG16 model
model = VGG16(weights='imagenet', include_top=True)
# Load and preprocess an example image
img = image.load_img('path_to_image.jpg', target_size=(224, 224))
img = image.img_to_array(img)
img = np.expand_dims(img, axis=0)
img = preprocess_input(img)
# Assume we are using 'output' as the final layer's output tensor
output = model.output[:, np.argmax(output)] # Replace with actual class
index if known
# Compute the gradients of the target class concerning the input image
grads = tf.gradients(output, model.input)[0]
# Normalize the gradients
pooled_grads = tf.reduce_mean(grads, axis=(0, 1, 2))
# Fetch the pooled gradients and the image for visualization
iterate = tf.keras.backend.function([model.input], [pooled_grads, grads])
pooled_grads_value, grads_value = iterate([img])
# Multiply each channel in the image array by "how important this
channel is" concerning the top predicted class
for i in range(pooled_grads_value.shape[-1]):
img[0, :, :, i] *= pooled_grads_value[i]
# Average over the channels to compute the saliency map
heatmap = np.mean(grads_value, axis=-1)[0]
```

```
# Visualize the heatmap
heatmap = np.maximum(heatmap, 0)
heatmap /= np.max(heatmap)
plt.matshow(heatmap)
plt.show()
# Overlay the heatmap on the original image
import cv2
original_img = cv2.imread('path_to_image.jpg')
heatmap = cv2.resize(heatmap, (original_img.shape[1], original_img.shape[0]))
heatmap = np.uint8(255 * heatmap)
heatmap = cv2.applyColorMap(heatmap, cv2.COLORMAP_JET)
superimposed_img = heatmap * 0.4 + original_img
# Save and display the image
cv2.imwrite('path_to_superimposed_image.jpg', superimposed_img)"
```

LIME, SHAP, AND INTEGRATED GRADIENTS

Understanding how a neural network arrives at a particular decision is pivotal, especially when these decisions have significant real-world consequences. As we focus on making these complex models more interpretable, three beginner-friendly approaches stand out for their ability to demystify decision-making. Let's look at them in detail now.

Local Interpretable Model-agnostic Explanations (LIME)

LIME is a technique that helps us understand and explain the predictions of any machine learning classifier without needing to know its inner workings. The core idea is to approximate the complex model locally with a simpler one that is interpretable, like a linear model.

SHapley Additive exPlanations (SHAP)

SHAP builds on the concept of game theory to explain the output of machine learning models. It assigns each feature an essential value for a particular prediction. This value is a Shapley value derived from cooperative game theory. It ensures a fair distribution of payoffs among players—here, the 'players' are the features.

Integrated Gradients for Pixel-wise Attributions

Integrated Gradients are often used with deep learning models, particularly for tasks like image recognition. It helps us understand which pixels in an image contribute most strongly to the model's decision. This method measures the gradient of the model's output concerning the input image.

When making neural networks more interpretable, you must first understand their black-box nature and why making them interpretable is necessary. It's important to remember that using visualization techniques to understand the model's behavior can help. In addition, you can also use feature attribution models and create saliency maps to make the model more interpretable. In the next chapter, we'll focus on how you can stay updated with all advancements related to neural networks.

REFERENCES

- https://towardsdatascience.com/why-we-will-never-open-deep-learnings-black-box-4c27cd335118
- https://www.nature.com/articles/d41586-022-00858-1
- https://www.frontiersin.org/articles/10.3389/fpsyt.2020.551299/full
- https://escholarship.org/uc/item/0sk7h5m7

* https://www.interpretable.ai/interpretability/why/
* https://machinelearningmastery.com/how-to-visualize-filters-and-feature-maps-in-convolutional-neural-networks/
* https://towardsdatascience.com/convolutional-neural-network-feature-map-and-filter-visualization-f75012a5a49c
* https://www.kaggle.com/code/arpitjain007/guide-to-visualize-filters-and-feature-maps-in-cnn
* https://proceedings.neurips.cc/paper/7875-visualizing-the-loss-landscape-of-neural-nets.pdf
* https://mathformachines.com/posts/visualizing-the-loss-landscape/
* https://openreview.net/pdf?id=Sy21R9JAW
* https://scholarworks.utrgv.edu/cgi/viewcontent.cgi?article=1024&context=ece_fac
* https://cgl.ethz.ch/Downloads/Publications/Papers/2019/Anc19c/Anc19c.pdf
* https://towardsdatascience.com/practical-guide-for-visualizing-cnns-using-saliency-maps-4d1c2e13aeca
* https://christophm.github.io/interpretable-ml-book/pixel-attribution.html
* https://distill.pub/2020/attribution-baselines
* https://c3.ai/glossary/data-science/lime-local-interpretable-model-agnostic-explanations/
* https://christophm.github.io/interpretable-ml-book/lime.html
* https://towardsdatascience.com/understanding-deep-learning-models-with-integrated-gradients-24ddce643dbf
* https://towardsdatascience.com/understanding-deep-learning-models-with-integrated-gradients-24ddce643dbf
* https://www.frontiersin.org/articles/10.3389/fpsyt.2020.551299/full
* https://link.springer.com/article/10.1007/s12559-023-10179-8
* https://www.interpretable.ai/interpretability/why/
* https://journals.sagepub.com/doi/10.1177/2053951715622512
* https://arxiv.org/abs/1712.09923
* https://machinelearningmastery.com/how-to-visualize-filters-and-feature-maps-in-convolutional-neural-networks/

- https://www.kaggle.com/code/arpitjain007/guide-to-visualize-filters-and-feature-maps-in-cnn

- https://www.sciencedirect.com/science/article/pii/S0893608098001166

- https://proceedings.neurips.cc/paper_files/paper/2018/file/a41b3bb3e6b050b6c9067c67f663b915-Paper.pdf

- https://debuggercafe.com/saliency-maps-in-convolutional-neural-networks/

- https://christophm.github.io/interpretable-ml-book/pixel-attribution.html

- https://datascientest.com/en/shap-what-is-it

- https://christophm.github.io/interpretable-ml-book/pixel-attribution.html

## 7

STAYING UPDATED WITH NEURAL NETWORK ADVANCEMENTS

"Let's go invent tomorrow instead of worrying about what happened yesterday."

— STEVE JOBS

In the rapidly advancing field of neural networks, staying updated is necessary. You need to understand the fundamental concepts and keep pace with the continuous flow of innovations and research breakthroughs. As a beginner, the prospect of keeping up with such a dynamic field might seem daunting.

With overwhelming information, identifying relevant and authoritative content can be challenging. It is essential to have a structured approach to filter through the noise and focus on what is imperative for your growth and understanding. Throughout this chapter, we will explore various resources tailored to those new to the field.

State-of-the-Art Performance

Staying at par with the latest developments isn't just an academic exercise; it directly impacts the efficacy of the neural network models you develop. The adoption of state-of-the-art techniques can significantly enhance performance metrics. This can help you reduce error rates in predictive models or accelerate the training time of complex networks.

Moreover, it positions you to make informed decisions about trade-offs in model complexity, training costs, and execution speed. By staying current, you are better equipped to push the boundaries of what's possible, leveraging leading-edge advancements to deliver robust, innovative solutions.

New Architectures and Models

The introduction of new neural network architectures is a catalyst for change. These architectures are often designed to address specific challenges or improve upon existing models' limitations. For instance, the advent of transformers has dramatically changed the landscape of natural language processing tasks.

By understanding these new architectures, you can expand your problem-solving toolkit. It is not about adding another tool; it's about understanding which tool is most appropriate for the task. Familiarizing yourself with these models allows you to tackle a broader spectrum of problems, such as image recognition and sequential data analysis.

NAVIGATING THE RAPIDLY EVOLVING FIELD NEURAL NETWORKS

In this field of neural networks, staying updated is advancing your career, driving innovation, and achievir excellence. Therefore, you need to learn how to na harness the latest breakthroughs and resources networks.

Whether a beginner or a seasoned practitioner, you r valuable strategies for leveraging state-of-the-art model that in mind, let's look at some factors you need to sta with to navigate the field of neural networks.

Rapid Advancements

A rapid pace of progression marks the domain of neural Innovations are a constant, with an unending stream of niques and algorithms emerging from research institu tech companies. The volume of research papers is reflecting the relentless quest for improvement and opt in the field.

As a professional in this domain, keeping track of these ments is essential. This means regular engagement with journals, attendance at industry conferences, and active p tion in relevant tech communities. Recognizing which ments are gaining traction can provide insight into futur and shifts in best practices.

Improved Techniques and Algorithms

Continuous improvements in techniques and algorithms are essential to improve the neural network's performance. The enhancement of these elements not only refines current practices but also pushes the boundaries of what these models can achieve. These updates and improvements can help introduce more efficient:

1. Training Algorithms

The evolution of training algorithms is at the heart of neural network advancement. In many applications, novel approaches such as adaptive learning rate algorithms like Adam and RMSprop have already replaced traditional stochastic gradient descent (SGD). This is because of their ability to adjust the learning rate during training dynamically.

This results in improved convergence rates and can significantly reduce training time. Furthermore, newer variations and improvements in these adaptive algorithms are continuously being developed, offering more nuanced control over the learning process and convergence.

2. Regularization Methods

Regularization methods are crucial to combat overfitting. Techniques like dropout, where random neurons are ignored during training, simplify the model and prevent the co-adaptation of feature detectors. L1 and L2 regularizations are also widely employed, adding a penalty for larger weights to the loss function.

These methods have been further refined, leading to more sophisticated approaches like elastic net regularization that combines L1 and L2 methods and batch normalization, which helps regularize and accelerate deep network training.

3. Optimization Techniques

Beyond the traditional methods, recent advancements have introduced concepts like Lookahead Optimizers, which periodically update the learning direction based on an extrapolated future step, and the incorporation of second-order information in optimization through quasi-Newton methods. These sophisticated techniques offer faster convergence and stability in training deep neural networks.

On the architectural level, strategies such as skip connections and attention mechanisms have drastically improved the training and performance of deep networks. Skip connections, as used in Residual Networks (ResNets), allow gradients to flow through networks more effectively. This helps facilitate the training of much deeper networks by mitigating the vanishing gradient problem.

In addition, attention mechanisms, particularly self-attention, allow networks to focus on specific parts of the input sequentially, improving performance on tasks requiring contextual awareness. It would be best if you remembered that a combination of advanced techniques and algorithms enhances the performance of neural networks.

Circumventing Outdated Practices

Another you need to do to navigate the field of neural networks is to sidestep outdated practices. This includes obsolete techniques, approaches, and libraries. It's important to know that such practices can decrease the efficiency of your AI models. Let's look at this in more detail.

1. Outdated Techniques

Early neural network architectures often relied heavily on fully connected layers prone to overfitting and excessive computational costs. In contrast, modern architectures favor convolutional layers for tasks like image recognition, which significantly reduce the number of trainable parameters without sacrificing the ability to capture essential features.

2. Deprecated Libraries:

Regarding software, relying on outdated libraries can reduce efficiency and support. An example can be seen in the transition from Theano to TensorFlow or PyTorch. While Theano was a pioneering library in neural network development, it is no longer maintained, making it a less viable option for current projects. Staying updated ensures you use supported, robust, and optimized tools for the latest hardware, like GPUs and TPUs.

3. Obsolete Approaches

Certain practices, like using Sigmoid activation functions in hidden layers, have fallen out of favor due to the problem of vanishing gradients they introduce, making training deep networks challenging. ReLU and its variants like Leaky ReLU and Parametric ReLU have become the standard, as they help mitigate this issue.

Understanding Limitations and Challenges

Keeping pace with current research also clarifies neural networks' limitations and challenges. You need to know that this can help you make informed decisions about developing neural networks. Let's look at both limitations and challenges in more detail.

1. Limitations

A prevalent limitation is the interpretability of deep learning models. While architectures have become more complex and powerful, understanding the reasoning behind their decisions remains elusive. For instance, while a convolutional neural network can excel at image classification, explaining the specific reasons for its classifications in human-understandable terms, often referred to as explainable AI (XAI), is still under active research.

2. Challenges

Another challenge is handling imbalanced datasets, where classes are not equally represented. Traditional neural network training methods can be biased towards the majority class, resulting in poor performance in minority classes. Techniques such as SMOTE (Synthetic Minority Over-sampling Technique) for oversampling the minority class or modified loss functions that emphasize the minority class are current solutions.

Staying informed about these evolving limitations and challenges not only aids in avoiding obsolete practices but also fosters a mindset geared toward innovation and problem-solving. By understanding where current approaches fall short, you can direct your efforts toward areas of research that can propel you further in the field of neural networks.

Leveraging New Tools And Frameworks

In a field as dynamic as neural network development, the advent of new tools, libraries, and frameworks can be transformative. They equip you with the means to enhance the development process, boost productivity, and unlock new capabilities. You need to be aware of advancements in tools, libraries, frameworks, and development processes.

1. New Tools And Libraries

Innovative tools like Netron allow developers to visualize neural network architectures, offering clarity on model structure that can aid optimization and debugging. Libraries such as Hugging Face's Transformers provide pre-trained models, a boon for natural language processing tasks, enabling state-of-the-art performance without extensive computational resources.

2. Advancements in Frameworks

On the framework front, TensorFlow and PyTorch continue to evolve. TensorFlow's latest versions have improved support for eager execution, which enhances debugging and dynamic computation graph generation. Meanwhile, PyTorch has gained popularity for its intuitive syntax and dynamic computation graph, which are particularly conducive to rapid prototyping and research.

Frameworks like Fast.ai build on top of these lower-level libraries to offer a higher-level abstraction, enabling developers to implement complex neural network architectures with less code. This can significantly reduce development time and make deep learning more accessible.

3. Streamlining Development Process:

Moreover, tools such as MLflow and DVC (Data Version Control) have emerged to manage the machine learning lifecycle, including experimentation, reproducibility, and deployment. MLflow tracks experiments and bundles libraries to streamline the transition from development to production. Conversely, DVC focuses on versioning datasets and machine learning models, facilitating better data management practices in the machine learning workflow.

4. Improving Productivity and Functionality

Additional functionalities these new tools offer include automated hyperparameter tuning, which is essential in optimizing neural network performance. Platforms like Weights & Biases and Ray Tune help track and fine-tune these parameters over numerous training runs, providing insights that guide the refinement process.

Leveraging these tools and frameworks effectively can lead to more robust neural network implementations. Therefore, you need to stay updated as it will allow you to harness these advancements. Ensuring they utilize the most efficient and powerful resources to develop neural network models.

Networking And Collaboration

In the field of neural networks, where the pace of innovation is relentless, networking and collaboration stand as pillars of personal and professional development. This means that you need to engage actively with the AI community. This can catapult your understanding and open up avenues for advancement and collaboration. Some of the ways you can network and collaborate with others include:

1. Conferences And Workshops

Conferences such as NeurIPS, ICML, and CVPR are essential for anyone serious about staying abreast of neural network advancements. These gatherings are not just for presenting breakthrough research; they offer workshops, tutorials, and symposia where one can learn directly from leading experts. Participation in these events provides a firsthand look at emerging trends and the opportunity to question, discuss, and understand the intricacies of the latest research.

2. Professional Discussions and Forums

Online forums and discussion groups like Reddit's r/MachineLearning, Cross Validated on Stack Exchange, and specific groups on LinkedIn or ResearchGate serve as valuable discussion and exchange platforms. Queries posed to these communities can yield a spectrum of insights, from practical advice on model implementation to theoretical discussions on algorithmic challenges.

3. Collaborative Platforms:

Platforms like GitHub are instrumental in collaboration. They are not just code repositories but serve as communities where one can contribute to open-source projects, inspect cutting-edge code, or even create collaborative projects that attract contributions from around the globe. Collaborating on projects through GitHub can expose developers to new coding practices, architectural strategies, and problem-solving approaches.

4. Knowledge Sharing Events

Local meetups and hackathons sponsored by AI interest groups and tech companies are fertile ground for collaboration. They can act as catalysts for forming study groups or initiating collaborative projects. For instance, the Deep Learning Indaba or the TensorFlow User Groups offer spaces to learn and contribute to communal knowledge.

5. Professional Growth Opportunities

Engaging with the community through these mediums leads to a cross-pollination of ideas, vital for innovation. Moreover, these interactions can often lead to professional opportunities such as joint research projects, consulting roles, or even job offers from organizations seeking expertise in neural networks.

By actively participating in the AI community, you're not just a spectator; you become a part of a broader conversation about neural networks. This connection means you'll keep learning from others, whether it's through online forums, project collaborations, or simply following the latest discussions and trends. It helps you stay in the loop with new techniques, tools, and practices shaping the future of neural networks. You can apply what you learn directly to your work, keeping your skills sharp and your projects innovative.

Innovating and Pushing Boundaries

Staying current in the neural network space is not just about adapting to new trends; it's about being part of the force that moves the field forward. You're better equipped to leverage the latest developments and create innovative solutions when you're up to date with the latest research and technology. There are several things you can do to foster innovation; some of them include:

1. Research Exploration

You need to know that exploring attention mechanisms led to the development of transformer models. By understanding and leveraging these mechanisms, researchers at Google introduced models like BERT and GPT, which have revolutionized natural language processing. These models emerged from a deep engagement with the latest findings and a willingness to venture into relatively uncharted territories of neural network design.

2. Developing Innovative Solutions

Innovations arise from applying neural networks to new problems. For example, AlphaFold by DeepMind demonstrated how neural networks could predict protein folding patterns, a problem considered intractable for decades. Staying informed about such advancements can inspire you to apply neural network technology to issues in your field that may benefit from a similar approach.

3. Pushing Technical Boundaries

On the technical side, there's ongoing research into reducing the computational cost of training large neural networks. Techniques like network pruning, quantization, and knowledge distillation result from focused efforts to make neural networks more efficient. By staying informed about these techniques, you could make neural networks more accessible and sustainable.

4. Exploring New Research Directions

Moreover, keeping track of the latest papers and preprints on platforms like arXiv can give you insight into the bleeding edge of neural network research. By engaging with these works, you can identify gaps in the current understanding and propose novel research directions. This can even help you develop a new perspective on adversarial training or uncover a way to improve generative models.

5. Developing Complex Problem Solutions

By incorporating the latest research into your work, you can develop solutions to complex problems that may have been unsolvable with older methods. For instance, recent advancements in reinforcement learning have enabled more sophisticated decision-making models. These models are now being applied in areas ranging from game playing to autonomous vehicle navigation.

Career Opportunities

Keeping up to date with the latest developments in neural networks is an investment in your professional future. The expertise you gain through continuous learning directly translates into expanded career opportunities across various sectors increasingly reliant on AI. Some of the factors you need to consider include:

1. Industry Demand

Companies in the tech industry are constantly looking for skilled professionals who can contribute to developing cutting-edge AI applications. For instance, Google's DeepMind regularly recruits experts capable of advancing AI research. The engineers and researchers working on AlphaFold possess a deep understanding of neural networks, which was paramount to their success in protein folding prediction.

2. Academic Contributions

Your up-to-date knowledge can position you as a candidate for research roles in prestigious institutions. Universities value researchers who bring fresh perspectives from industry trends into their academic endeavors. For example, researchers in different institutions are exploring areas like neuro-symbolic AI, which combines neural network approaches with symbolic reasoning, opening up new paths in AI research.

3. Enhancing Professional Value

Your value as an AI practitioner increases significantly when you're well-versed in the latest technologies and methodologies.

For example, knowledge of state-of-the-art models like GPT-4 or the latest breakthroughs in computer vision using convolutional neural networks can set you apart in a field that's becoming increasingly competitive.

4. Innovation in Non-Tech Industries

The demand for AI expertise extends beyond the tech industry. Sectors such as healthcare, finance, and automotive are seeking professionals who can apply neural network innovations to their specific challenges. For example, practitioners who understand how to apply neural networks to medical imaging analysis for disease diagnosis are in high demand in healthcare.

5. Research Opportunities

Being updated also positions you well to participate in grant-funded research projects requiring the latest AI knowledge. Agencies like the National Science Foundation (NSF) and the European Research Council (ERC) fund projects where neural network expertise can significantly impact. This includes things like improving climate models or enhancing cybersecurity with AI.

By maintaining a current understanding of neural network advancements, you secure your relevance in the job market and enhance your potential for higher earnings and positions of influence. Mastering current AI technologies, particularly neural networks, thus becomes a pivotal factor in shaping your career trajectory in research and applied AI fields.

BEGINNER-FRIENDLY RESOURCES FOR STAYING UPDATE

In the rapidly evolving landscape of neural networks, staying updated with the latest developments is a continuous challenge. You can use specific online platforms that provide invaluable resources to remain informed and enhance your expertise. This includes blogs, forums, tutorial websites, and conferences. Let's briefly go over them now.

Blogs

Online blogs are a great way to gain in-depth and up-to-date knowledge in a short amount of time. Some of the best blogs include:

- **Towards Data Science** - a hub of shared knowledge and expertise from data science professionals and enthusiasts. Here, you can find articles on cutting-edge neural network techniques and practical tips for implementation.
- **Medium** - a platform where thought leaders and innovators discuss the latest trends, tools, and research findings in neural networks and broader AI topics.
- **Distill** - an online platform for those who appreciate the intersection of art and science. It publishes clear, visually intuitive explanations of complex concepts in machine learning, often with interactive visualizations.
- **Sebastian Ruder's Blog** - offers in-depth discussions and analyses of the latest research and developments in natural language processing, a subfield of neural networks.

Forums and Communities

Forums and communities are an excellent way to share your perspective and learn from others. Some of the best forums and communities include:

- **MachineLearning subreddit** - a vibrant community for discussing the latest in machine learning and neural networks, where researchers and practitioners share insights and seek advice.
- **Kaggle Forums** - part of a larger data science competition platform and is ideal for those looking to apply their neural network knowledge to real-world problems and learn from others' approaches.
- **AI Stack Exchange** - a question-and-answer site for people interested in conceptual questions about life and challenges in a world where "cognitive" functions can be mimicked in purely digital environments.

Websites and Newsletters

This is a feasible option for those who like to receive tailored content. Some competent online websites and newsletters include:

- **arXiv** - repository of e-prints approved for publication after moderation but not full peer review. It is rich with the latest research papers on neural networks before they are formally published.
- **OpenAI Blog** - provides insights and updates on the cutting-edge work by OpenAI researchers, often providing early access to groundbreaking advancements.

- **Google AI Blog and DeepMind Blog** - platform where Google and its AI subsidiary DeepMind post updates on their research and discoveries, from new neural network architectures to novel applications.
- **NVIDIA Developer Blog** - platform for resources and discussions about the hardware that powers a lot of neural network computations, emphasizing GPUs and their software ecosystem.

In addition to these online resources, attending conferences such as NeurIPS, ICML, and ICLR can be game-changing. They present opportunities to hear about the latest research, network with professionals, and even present your work. These conferences often offer online access to papers and presentations for those who cannot attend in person.

As you journey through the dynamic world of neural networks, remember that staying informed is critical. By utilizing a combination of scholarly articles, collaborative forums, and pioneering conferences, you keep your skills sharp and your knowledge fresh.

Embrace these resources as your toolkit to navigate the nuances of neural network advancements. This can help you develop innovative solutions, career growth, and a deeper understanding of the transformative power of AI.

When trying to stay updated, remember to consider different factors such as career opportunities, networking, frameworks, and technologies and algorithms. And that's all you need to know to stay updated with neural network advancement. In the next chapter, we'll discuss beginner-friendly projects you can start with.

REFERENCES

- https://direct.mit.edu/neco/article/34/6/1289/110645/Advancements-in-Algorithms-and-Neuromorphic
- https://www.geeksforgeeks.org/neural-network-advances/
- https://www.weforum.org/agenda/2023/02/experts-ai-developing-over-the-coming-years
- https://direct.mit.edu/neco/article/34/6/1289/110645/Advancements-in-Algorithms-and-Neuromorphic
- https://medium.com/analytics-vidhya/a-complete-guide-to-adam-and-rmsprop-optimizer-75f4502d83be
- https://towardsdatascience.com/l1-and-l2-regularization-methods-ce25e7fc831c
- https://paperswithcode.com/method/lookahead
- https://towardsdatascience.com/bfgs-in-a-nutshell-an-introduction-to-quasi-newton-methods-21b0e13ee504
- https://pypi.org/project/Theano/
- https://builtin.com/machine-learning/relu-activation-function
- https://www.ibm.com/topics/explainable-ai
- https://medium.com/@corymaklin/synthetic-minority-over-sampling-technique-smote-7d419696b88c
- https://netraneupane.medium.com/netron-a-visualizer-for-machine-learning-and-deep-learning-models-470ad3a591
- https://www.simplilearn.com/tutorials/deep-learning-tutorial/deep-learning-frameworks
- https://medium.com/@haythemtellili/optimization-of-the-life-cycle-of-an-ml-project-using-mlflow-and-dvc-646553985ca0
- https://docs.ray.io/en/latest/tune/examples/tune-wandb.html
- https://www.am.ai/en/blog/ai-conferences-2023/
- https://www.reddit.com/r/MachineLearning/
- https://deepmind.google/technologies/alphafold/

- https://towardsdatascience.com/&sa=D&source=docs&ust=1699597596842191&usg=
AOvVaw1WY5VUUfZgDxLFxCWbRoNH
- https://medium.com/&sa=D&source=docs&ust=1699597609540758&usg=
AOvVaw0Dm9He7U4qPlFX3JJ60_CC
- https://distill.pub/
- https://www.ruder.io/
- https://www.kaggle.com/forums
- https://ai.stackexchange.com/
- https://ai.stackexchange.com/
- https://openai.com/blog/
- https://openai.com/blog/
- https://deepmind.com/blog/
- https://deepmind.com/blog/

## 8

PUTTING IT ALL TOGETHER: BEGINNER-FRIENDLY PROJECTS

"Innovation is the outcome of a habit, not a random act."

— SUKANT RATNAKAR

Now that we've reached the end of the book let's go on a journey through the world of neural networks. In this chapter, we'll look at some beginner-friendly projects you can try out on your own. We'll cover everything from data preprocessing to the actual code itself. So, Let's get started.

APPLYING NEURAL NETWORKS TO BEGINNER-FRIENDLY REAL-WORLD PROJECTS

In this section, we'll delve into practical applications of neural networks through beginner-friendly projects. These hands-on projects will equip you with valuable experience and insights into deep learning. Let's explore some exciting ideas where you can test your knowledge and apply neural networks effectively.

Digit Recognition Using The MNIST Dataset

Digit recognition is a classic problem in machine learning and a great starting point for understanding neural networks. The MNIST dataset, which consists of handwritten digits, is a commonly used benchmark for this task. It plays a fundamental role in benchmarking and testing various image classification algorithms, particularly those based on neural networks.

The MNIST dataset consists of a total of 70,000 handwritten digits, which are divided into two main subsets. These subsets include the training data set, which has 60,000 examples, and the test dataset, which has 10,000. The dataset covers all ten digits (0-9), making it a multi-class classification problem.

You need to know that each digit in the dataset is represented as a 28x28-pixel grayscale image. This means that each image has 28 rows and 28 columns, and each pixel's value represents the darkness of that pixel (0 for white, 255 for black). Despite its simplicity, MNIST remains relevant in machine learning education and research. It serves as a stepping stone for beginners to understand the concepts of data preprocessing, model architecture, training, and evaluation in the context of neural networks.

MNIST-like datasets have also been created for more complex tasks, such as the Fashion MNIST dataset for clothing classification. By working on this project, you'll get to grips with the basics of neural network architecture and image classification. To start working on this beginner-friendly project, follow a series of steps. These steps include:

1. **Data Preparation** - You must preprocess and load the MNIST dataset, ensuring it's ready for training and evaluation.
2. **Model Architecture** - Once done, you must design a neural network architecture suitable for image classification, typically a convolutional neural network (CNN).
3. **Training** - Then, you can train your model using the training data, adjusting hyperparameters and optimizing as needed.
4. **Evaluation** - After that, Assess the model's performance on the test dataset to gauge its accuracy in digit recognition.
5. **Fine-tuning** - Once the evaluation is complete, you can experiment with different network architectures, regularization techniques, and hyperparameters to improve your model's performance.

You'll need to write code to perform digit recognition using the MNIST dataset in Python. Here's an outline of the code you would typically use for this project using popular deep-learning libraries like TensorFlow and Keras. This code provides a high-level overview of the process.

The code loads the MNIST dataset, preprocesses it, builds a CNN model, compiles it, trains the model on the training data, and evaluates its accuracy on the test data.

"import tensorflow as tf
from tensorflow.keras.datasets import mnist
from tensorflow.keras.models import Sequential

```
from tensorflow.keras.layers import Dense, Conv2D, Flatten,
MaxPooling2D
from tensorflow.keras.utils import to_categorical
# Load MNIST dataset
(train_images, train_labels), (test_images, test_labels) = mnist.load_data()
# Normalize the images.
train_images = train_images / 255.0
test_images = test_images / 255.0
# Reshape dataset to have a single channel
train_images = train_images.reshape((train_images.shape[0], 28, 28, 1))
test_images = test_images.reshape((test_images.shape[0], 28, 28, 1))
# Convert labels to one-hot vectors
train_labels = to_categorical(train_labels)
test_labels = to_categorical(test_labels)
# Build the model
model = Sequential([
Conv2D(32, kernel_size=3, activation='relu', input_shape=(28, 28, 1)),
MaxPooling2D(pool_size=2),
Flatten(),
Dense(128, activation='relu'),
Dense(10, activation='softmax')
])
# Compile the model
model.compile(optimizer='adam', loss='categorical_crossentropy',
metrics=['accuracy'])
# Train the model
model.fit(train_images, train_labels, validation_data=(test_images,
test_labels), epochs=10)
# Evaluate the model
loss, accuracy = model.evaluate(test_images, test_labels)
print(f'Loss: {loss}, Accuracy: {accuracy}')"
```

This code provides a basic yet comprehensive approach to digit recognition using neural networks. It's a great starting point for beginners to understand and experiment with neural network architectures in Python. Now, let's move on to another beginner-friendly project.

Sentiment Analysis On Movie Reviews

Sentiment analysis is used in natural language processing (NLP) to determine the emotional tone behind a body of text. This powerful tool helps understand the sentiments, opinions, and attitudes expressed in written language. A typical application of sentiment analysis is analyzing movie reviews to determine whether the feeling is positive, negative, or neutral.

When working on this project, the first step is gathering a movie review dataset. You can use public datasets, like the IMDb dataset, for this purpose. Once you have done that, you need to process the data following different steps that include:

1. **Text Cleaning** - To clean the text, you must remove unnecessary characters, symbols, and formatting.
2. **Tokenization** - You need to break the text into words or phrases after that.
3. **Stop Words Removal** - You need to eliminate common words with no significant sentiment. Common examples of such words include "is, this, that, the," and so on.
4. **Stemming/Lemmatization:** The next thing you need to do is reduce words to their root form.
5. **Vectorization:** Lastly, you need to convert text into a numerical format (like TF-IDF or Word Embeddings) that neural networks understand.

Once you're done with this, the next thing you need to do is build a neural network for sentiment analysis. This process will involve four steps that are mentioned below:

1. Input Representation - During this step, you need to make sure that the Input data should be transformed into fixed-size vectors. This can be achieved using techniques like Bag of Words, TF-IDF, or Word Embeddings (Word2Vec or GloVe).

2. Model Architecture - Next, you must create a model architecture. Here, you have three options that include:

a. Simple Feedforward Neural Network - A basic approach, though less practical for capturing text contextual information.

b. Recurrent Neural Networks (RNNs) - More advanced, capable of handling sequences and understanding text context. LSTM (Long Short-Term Memory) or GRU (Gated Recurrent Units) are popular choices.

c. Embedding Layer - Used at the input layer to convert word indices into fixed-sized dense vectors.

3. Training The Model - To train the model effectively, you must use binary cross-entropy as the loss function for binary classification (positive/negative).

4. Model Evaluation and Improvement - You need to use metrics like accuracy, precision, recall, and F1-score to evaluate the model. You can also experiment with network architectures, hyperparameters, and embedding techniques.

Here's an example of implementing a sentiment analysis model for movie reviews using Python. This example will use TensorFlow and Keras, and the IMDb dataset is conveniently available through

Keras. We'll use a simple LSTM (Long Short-Term Memory) model, a recurrent neural network (RNN) suitable for dealing with sequences like text.

```
"import tensorflow as tf
from tensorflow.keras.datasets import imdb
from tensorflow.keras.preprocessing.sequence import pad_sequences
from tensorflow.keras.models import Sequential
from tensorflow.keras.layers import Embedding, LSTM, Dense
# Load the IMDb dataset
max_features = 10000 # Number of words to consider as features
maxlen = 500 # Cuts off reviews after 500 words
(x_train, y_train), (x_test, y_test) = imdb.load_data(num_-
words=max_features)
# Preprocess the data: Pad sequences
x_train = pad_sequences(x_train, maxlen=maxlen)
x_test = pad_sequences(x_test, maxlen=maxlen)
# Build the model
model = Sequential()
model.add(Embedding(max_features, 128, input_length=maxlen))
model.add(LSTM(32))
model.add(Dense(1, activation='sigmoid'))
# Compile the model
model.compile(optimizer='adam', loss='binary_crossentropy', metrics=
['accuracy'])
# Train the model
model.fit(x_train, y_train, epochs=5, batch_size=32, validation_s-
plit=0.2)
# Evaluate the model
loss, accuracy = model.evaluate(x_test, y_test)
print(f'Test Loss: {loss}, Test Accuracy: {accuracy}')"
```

In this code, the IMDb dataset is loaded from Keras and is limited to the top 10,000 most frequently used words. It is then preprocessed, where sequences are padded to ensure they have the same length. An Embedding layer is the first layer to turn positive integers (indexes) into fixed-sized dense vectors.

An LSTM layer is added to process sequences. The output layer is a Dense layer with a sigmoid activation function for binary classification. The model is compiled with the binary cross-entropy loss function and the Adam optimizer. Once the compilation is complete, the model is trained on the training data for a defined number of epochs. Lastly, the model is evaluated on the test data to determine its accuracy. Now, let's look at another beginner-friendly project you can try.

Predicting Stock Prices Using Historical Data

Predicting stock prices is a challenging yet fascinating application of neural networks. It involves analyzing historical data on stock prices to predict future trends. This task falls under the category of time series forecasting, a domain where neural networks, especially recurrent neural networks (RNNs) like LSTM (Long Short-Term Memory), excel due to their ability to capture temporal dependencies.

Stock market data is typically time-series data comprising various attributes like opening price, closing price, highest price of the day, lowest price, and trading volume. For beginners, focusing on one variable, like the closing price, is a practical approach to simplify the problem. To preprocess the data for this project, there are two things you need to do:

- **Normalization** - Stock price data should be normalized, often by scaling the data to a range of 0 to 1. This helps speed up the training process and reduce the likelihood of difficulties due to numerical instability.
- **Windowing** - Transform the time series data into a supervised learning problem. Data is arranged in windows representing the input features (past prices) and the output label (future price).

When you're building a neural network, you need three things: the model's architecture, training, and evaluation. As far as the model's architecture is concerned, the LSTM layers and the Dense layer are critical. The LSTM layers are crucial for capturing the long-term dependencies in time-series data.

Meanwhile, the Dense layer is crucial for capturing the long-term dependencies in time-series data. When training the model, you can use Mean Squared Error (MSE) as the loss function, which is typical for regression problems. In addition, an optimizer like Adam or RMSprop is appropriate for these tasks.

You can then evaluate the model using metrics like Mean Absolute Error (MAE) or Root Mean Squared Error (RMSE). However, assessing the model on unseen data is essential to understand its predictive power.

Below is an example of how you might implement a simple neural network in Python to predict stock prices using historical data. This example uses LSTM (Long Short-Term Memory), a recurrent neural network suitable for time series data. For demonstration purposes, we'll assume the historical stock data is in a CSV file and includes a 'Close' column for the closing prices.

```
"import numpy as np
import pandas as pd
from tensorflow.keras.models import Sequential
from tensorflow.keras.layers import LSTM, Dense
from sklearn.preprocessing import MinMaxScaler
from sklearn.model_selection import train_test_split
# Load the dataset
df = pd.read_csv('stock_prices.csv')
data = df['Close'].values
data = data.reshape(-1, 1)
# Normalize the data
scaler = MinMaxScaler(feature_range=(0, 1))
data = scaler.fit_transform(data)
# Prepare the dataset for LSTM
def create_dataset(data, time_step=1):
X, y = [], []
for i in range(len(data)-time_step-1):
a = data[i:(i+time_step), 0]
X.append(a)
y.append(data[i + time_step, 0])
return np.array(X), np.array(y)
time_step = 100
X, y = create_dataset(data, time_step)
X_train, X_test, y_train, y_test = train_test_split(X, y, test_size=0.2)
# Reshape input to be [samples, time steps, features] for LSTM
X_train = X_train.reshape(X_train.shape[0],X_train.shape[1], 1)
X_test = X_test.reshape(X_test.shape[0],X_test.shape[1], 1)
# Build the LSTM model
model = Sequential()
model.add(LSTM(50, return_sequences=True, input_shape=(100, 1)))
model.add(LSTM(50, return_sequences=False))
model.add(Dense(25))
```

```
model.add(Dense(1))
# Compile the model
model.compile(optimizer='adam', loss='mean_squared_error')
# Train the model
model.fit(X_train, y_train, batch_size=1, epochs=1)
# Predict and inverse transform the normalization
train_predict = model.predict(X_train)
test_predict = model.predict(X_test)
train_predict = scaler.inverse_transform(train_predict)
test_predict = scaler.inverse_transform(test_predict)
# Evaluate the model
test_loss = model.evaluate(X_test, y_test)
print(f'Test Loss: {test_loss}')"
```

This code provides a foundational approach to predicting stock prices with LSTM in Python. It's important to remember that real-world financial data prediction is far more complex and requires thorough analysis and consideration of many external factors.

PRACTICAL CHALLENGES AND TIPS FOR DEPLOYING MODELS

Model deployment is a critical step in the life cycle of any machine learning project. It marks the transition from theoretical models and controlled environments to real-world applications. This leap can seem daunting for beginners, with several technical and practical challenges to navigate.

Model Deployment And Serving Predictions

Deployment is not just about making a model; it's about integrating it into an environment where it can start making decisions based on new data. This could be a website, an app, or a backend server. The choice of where and how to deploy a model hinges on the needs of the project and the available resources.

For many, starting with a local server setup or a cloud-based platform like AWS, Google Cloud, or Azure is a practical choice. These platforms offer scalability and robustness, crucial for models that handle varied loads and complex computations. The real test of a deployed model is in its ability to serve predictions accurately and efficiently.

Tools like TensorFlow Serving, Flask, and Docker containers have become famous for their efficiency in managing and serving models. It's essential to have a setup that allows the model to receive input data, process it, and reliably return predictions.

Navigating Scalability And Performance Issues

One of the most significant challenges in deploying neural network models is ensuring they perform well when scaled up to handle larger datasets or more complex tasks than they encountered during training. Scalability is vital for the long-term success of a model, especially in dynamic environments like stock price prediction or real-time sentiment analysis.

To achieve this scalability, various strategies come into play. Data normalization or standardization is a critical step, ensuring that the model is fed consistent data within a scale that it can handle efficiently. Techniques like batch processing and parallel processing can also greatly enhance the performance of a model under load. Additionally, utilizing hardware acceleration through GPUs or TPUs can be a game changer for computationally intensive models.

However, even with these technical strategies in place, the unpredictable nature of real-world data and scenarios can pose challenges. Here, continuous monitoring, updating, and tweaking of the model become necessary. Performance metrics such as accuracy, precision, and recall offer insights into how the model is performing. Still, it's also crucial to be vigilant for signs of model drift or data changes that could impact performance.

Now that you know some beginner-friendly projects and know how to deploy models, go ahead and build your own project. Good Luck!

REFERENCES

- https://machinelearningmastery.com/how-to-develop-a-convolutional-neural-network-from-scratch-for-mnist-handwritten-digit-classification/
- https://www.analyticsvidhya.com/blog/2021/11/newbies-deep-learning-project-to-recognize-handwritten-digit/
- https://cs224d.stanford.edu/reports/PouransariHadi.pdf
- https://machinelearningmastery.com/predict-sentiment-moviereviews-using-deep-learning/
- https://www.google.com/url?q=https://learner-cares.medium.com/handwritten-digit-recognition-using-convolutional-neural-network-cnn-with-tensorflow-2f444e6c4c31
- https://machinelearningmastery.com/how-to-develop-a-convolutional-neural-network-from-scratch-for-mnist-handwritten-digit-classification/
- https://www.oracle.com/pk/artificial-intelligence/what-is-natural-language-processing/
- https://developer.imdb.com/non-commercial-datasets/
- https://speakai.co/word2vec-vs-glove/
- https://www.analyticsvidhya.com/blog/2021/03/introduction-to-long-short-term-memory-lstm/
- https://blog.marketmuse.com/glossary/gated-recurrent-unit-gru-definition/

- https://www.analyticsvidhya.com/blog/2022/03/a-brief-overview-of-recurrent-neural-networks-rnn/
- https://learn.microsoft.com/en-us/office/troubleshoot/access/database-normalization-description
- https://statisticsbyjim.com/regression/mean-squared-error-mse/
- https://deepchecks.com/glossary/mean-absolute-error/
- https://www.simplilearn.com/normalization-vs-standardization-article
- https://towardsdatascience.com/predicting-stock-prices-usinga-keras-lstm-model-4225457f0233
- https://journalofbigdata.springeropen.com/articles/10.1186/s40537-020-00333-6
- https://neptune.ai/blog/ml-model-serving-best-tools
- https://towardsdatascience.com/3-ways-to-deploy-machine-learningmodels-in-production-cdba15b00e
- https://machinelearningmastery.com/how-to-improve-neural-networkstability-and-modeling-performance-with-data-scaling/
- https://www.codementor.io/blog/scaling-ml-6ruo1wykxf

CONCLUSION

Congratulations are in order because you've managed to complete this book. Throughout the book, we have been focused on gaining an in-depth understanding of neural networks and their practical implementation in Python. The most important thing you must remember is that a deep knowledge of neural networks, coupled with practical skills in Python, is essential.

It empowers professionals in technology and research fields to innovate and excel. We started our journey by building a solid foundation of what neural networks are. This chapter set the foundation by providing information about neural networks, their real-world applications, natural language processing, and more.

We discussed that neural networks have found widespread application across various industries, revolutionizing how we approach problems and innovate. In healthcare, they're used for disease diagnosis and drug discovery, enhancing the accuracy and speed of medical interventions.

The finance sector leverages neural networks for fraud detection and algorithmic trading, improving security and financial decision-making. In technology, they drive advancements in natural language processing and image recognition, powering virtual assistants and facial recognition systems that have become integral to our daily lives.

In automotive industries, neural networks are crucial for developing self-driving car technologies and enhancing safety and efficiency. Additionally, they play a significant role in environmental modeling, helping predict climate patterns and natural disasters, thus contributing to proactive environmental management.

These diverse applications showcase neural networks' versatility and transformative potential in addressing complex, real-world challenges. Then, we shifted our focus to the math behind neural networks, making concepts like gradient descent and backpropagation accessible.

In this chapter, we learned the math behind neural networks, linear algebra for matrix operations, and calculus for derivatives and gradients. Before we shifted our focus to implementing neural networks, we discussed that it's essential to introduce gradient descent and backpropagation algorithms.

Regarding the implementation, we learned about building, testing, and evaluating neural networks in Python. It would be best if you remembered that you should use tools like Anaconda and other necessary libraries to set up a virtual environment in Python. Handling the intricacies of neural networks, in the latter chapters, we explored advanced topics, equipping you with the tools to navigate the complex world of neural networks.

By addressing these issues, you can increase your neural network development capabilities and handle various situations in future projects. Then, we covered the critical aspect of data preparation for neural networks. This chapter provided you with an in-depth understanding of methodologies for preprocessing and transforming data, which is crucial for the success of your neural network models.

We covered practical techniques like handling missing data, scaling features, and applying data augmentation strategies. It's important to remember that all of these were aimed at improving the quality and reliability of your training process. Handling missing data, for instance, is a common challenge in dataset preparation.

By applying methods to fill or exclude missing information intelligently, you can significantly enhance the integrity of your data, leading to more robust model performance. Similarly, feature scaling, an often overlooked yet critical step, involves adjusting the range of your data features to a standard scale.

This standardization ensures that your neural network doesn't bias towards certain features over others due to the sheer difference in scale, thereby improving model accuracy and efficiency. Another technique we discussed was data augmentation. This strategy is beneficial in scenarios where data is scarce or too uniform.

You can expand your dataset by artificially creating variations of your existing data, such as rotating images or altering the pitch in audio samples. This enables your neural network to learn from a broader range of examples and thus become more generalized and less prone to overfitting.

This chapter equipped you with tools to build, interpret, and explain the decisions made by your models. Remember, with visualization techniques, feature importance methods, and beginner-friendly explainable AI approaches, you can ensure transparency and accountability in your AI applications.

By visualizing how different network layers respond to various inputs, you can gain insights into what the model sees and learns. Techniques like heat maps for CNNs can show which parts of an input image are triggering the most robust responses in the network, helping you understand why a network might be making certain decisions.

This is particularly useful in applications like medical image analysis, where understanding the focus areas of a model can correlate strongly with its diagnostic decisions. Techniques like permutation feature importance or SHAP (SHapley Additive exPlanations) values allow you to quantify the impact of each feature on the model's output.

This not only aids in model interpretability but also guides you in feature selection and data preprocessing. This ensures that your model focuses on relevant information and reduces the risk of learning from noise or biased data. However, keeping pace with the rapid advancements in neural networks is essential.

You can use different resources to stay on par with the latest developments, including blogs, tutorials, research papers, and frameworks. However, you need to develop a habit of continuous learning. Lastly, we covered beginner-friendly projects like digit recognition, sentiment analysis, and stock price prediction, where you applied your accumulated knowledge in practical, impactful ways.

This hands-on experience is crucial in solidifying the concepts learned throughout the book, empowering you to confidently build and deploy your neural network models.

The approach of this book in explaining neural networks in Python is designed to have a lasting impact on you as a reader. It's not merely about grasping the current state of technology; it's about actively preparing you to contribute to its future. The skills and knowledge you've gained from this book are tailored to be directly applicable in various fields.

Whether you're engaged in academic research or embarking on industry projects, what you've learned will be of help. This knowledge is your toolkit for innovation and contribution to the ever-evolving landscape of technology. You need to use it and revolutionize how you work with Python, AI, and tech.

But before you head out and start creating neural networks, leave a review and mention what you liked about this book.

Good luck with your future projects!

REFERENCES

- https://www.analyticssteps.com/blogs/8-applications-neural-networks
- https://builtin.com/data-science/gradient-descent
- https://builtin.com/machine-learning/backpropagation-neural-network
- https://www.arch.jhu.edu/python-virtual-environments/
- https://towardsdatascience.com/7-ways-to-handle-missing-values-in-machine-learning-1a6326adf79e
- https://towardsdatascience.com/convolutional-neural-network-feature-map-and-filter-visualization-f75012a5a49c
- https://help.qlik.com/en-US/cloud-services/Subsystems/Hub/Content/Sense_Hub/AutoML/shap-importance.htm
- https://machinelearningmastery.com/how-to-develop-a-convolutional-neural-network-from-scratch-for-mnist-handwritten-digit-classification/
- https://machinelearningmastery.com/predict-sentiment-movie-reviews-using-deep-learning/
- https://towardsdatascience.com/predicting-stock-prices-using-a-keras-lstm-model-4225457f0233

www.ingramcontent.com/pod-product-compliance
Lightning Source LLC
Chambersburg PA
CBHW071414210326
41597CB00020B/3498